THE IDENTIFICATION OF FIREARMS AND FORENSIC BALLISTICS

Wolfe Library Classics Limited Edition

This is Copy No. 1305 *of 1500.*

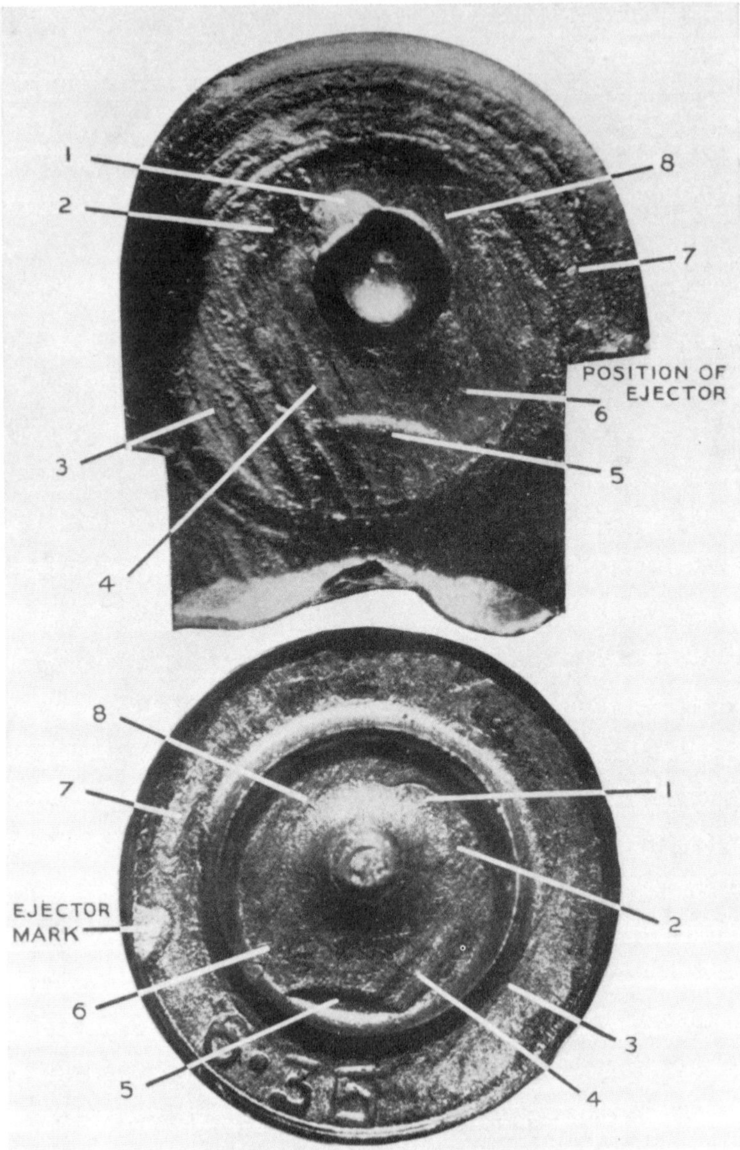

THE BREECH FACE OF A 6·35 MM. SELF-LOADING PISTOL AND THE BASE OF A CARTRIDGE WHICH WAS FIRED FROM THIS PISTOL

The marks on the breech face have been imprinted on the base of the cartridge, and in order to simplify the identification of the different marks and their corresponding imprints they have all been numbered. The following marks are indicated on the breech face: 1, a deep indentation possibly caused by an accidental blow with the end of a cleaning rod; 2, irregular roughening; 3, large tool mark in the shape of a segment of a circle; 4, large parallel tool cuts; 5, deep indentation possibly caused by an accidental blow with the end of a cleaning rod; 6, small indentation; 7, circular tool mark; 8, fine parallel tool marks. All these marks, together with that made by the ejector, have been imprinted on the base of the cartridge as indicated by the corresponding numbers

THE IDENTIFICATION OF FIREARMS AND FORENSIC BALLISTICS

BY

MAJOR GERALD BURRARD

D.S.O., R.F.A. (RETIRED)

Wolfe Publishing Company

6471 Airpark Drive

Prescott, Arizona 86301

Originally published by Butler & Tanner Ltd., 1934

Manufactured in the United States of America
Reprinted August 1990

ISBN: 0-935-632-91-3

1990
Wolfe Publishing Company
Prescott, Arizona

INTRODUCTION

DURING the past few years it has so happened that I have been called as an expert witness or adviser in various legal cases in which firearms played an important part. The experience which I have so gained has convinced me of the widespread ignorance of firearms which exists in quarters in which a correct understanding of the elementary principles of the subject would appear to be essential. In fact, all too frequently this ignorance can be described only as appalling.

I feel, therefore, that there is real want of a book which gives the fundamental essentials of firearms and ballistics which are likely to be of forensic importance. And feeling as I do, I have had the temerity to try to fill the blank.

The first part of this book is confined solely to forensic ballistics. The facts which I have given are elementary, but I know of no other source where a lawyer could discover them without an immense amount of search and reading.

The second part deals with the problem of the identification of individual firearms by means of microscopic examinations of fired bullets and fired cartridge cases. This is a subject in which I have been peculiarly interested during the past five years, and to which I have devoted a great deal of time and study. The procedure which I have suggested and the conclusions which I have reached are the outcome of my own work and personal experience. If they differ from those of other investigators, I can but say that I have accepted no result which I have not checked, re-checked and checked again.

Thanks to my friendship with Mr. R. K. Wilson, F.R.C.S., I have had exceptional opportunities for examining and testing various makes of pistols and revolvers which are rarely encountered in this country. Mr. Wilson's knowledge and collection have been unreservedly at my disposal; and it is a pleasure to place on record my indebtedness to him.

I owe much to my friend, Mr. C. F. Hill, who is not only recognized as one of the first masters of photomicrographic technique in this country, but who is equally well known as a long range match rifle shot at Bisley. My friend, Mr. F. W. Jones, O.B.E., has once again had the kindness to read through the drafts of this book; while for many years he has given me ungrudging help and kindly encouragement.

I have also received the most ready help from Messrs. W. Watson & Sons, who have spared neither time nor trouble in making up apparatus for my requirements, frequently at very short notice. I could not possibly have done my work without their generous co-operation.

Details of weapons and ammunition which can readily be obtained elsewhere have been omitted purposely, as their inclusion seemed redundant.

G. BURRARD.

WILLOW LODGE,
HUNGERFORD,
BERKS.
April, 1934.

CONTENTS

ILLUSTRATIONS

PLATES

LINE DRAWINGS

THE IDENTIFICATION OF FIREARMS AND FORENSIC BALLISTICS

THE IDENTIFICATION OF FIRE-ARMS AND FORENSIC BALLISTICS

CHAPTER I

TYPES OF FIREARMS

FIREARMS which are used in crimes of violence can be divided into two main types: Smooth-Bores and Rifled Arms. Both these types, especially the latter, can be subdivided into a number of varieties; but the forensic problems which arise will generally fall into two main categories corresponding with the two main types, although there will sometimes be a certain amount of overlapping. It will, therefore, be convenient first to consider the two main types of firearms, and then the various subdivisions, as a thorough understanding of these apparently elementary rudiments will smooth the way to appreciating and investigating the more difficult and technical questions which inevitably arise in crime detection.

Smooth-Bores.—A Smooth-Bore is a firearm in which the bore, or inside of the barrel, is perfectly smooth from end to end. The cross section at any point in the barrel will give a perfect circle for the bore.

Rifled Arms.—In all rifled arms the bore is cut longitudinally with a number of grooves. These grooves are parallel one to the other, but are cut with a twist, or spiral, from breech to muzzle. The cross-section of the barrel of a rifled arm presents an appearance of the type shown in Fig. 1.

The grooves in the bore are actually termed **grooves,** and those portions of the bore which are situated between the grooves are known as **lands.**

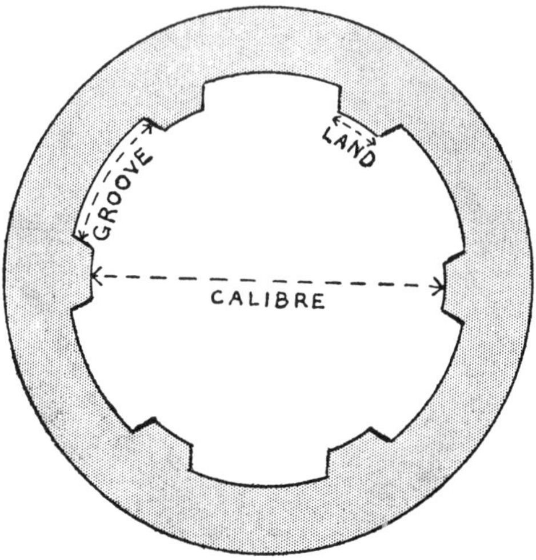

FIG. 1.—Cross-section of a rifled barrel showing the grooves and lands.

Gauge or Calibre.—Both types of firearms are classified by the Gauge, or Calibre, of their bores; that is by the internal dimension of the barrel. A different system of measuring, however, exists for large smooth-bores and rifled arms.

Large smooth-bores are measured by the number of spherical balls of pure lead, each exactly fitting the inside of the bore, which go to make up a pound. For example, a 12-bore (or 12-gauge) is a weapon in which the bore is of such a size that twelve spherical lead bullets, each exactly fitting the bore, will together weigh one pound.

This method of indicating the size of the bore is very old and dates back to the days of muzzle-loading cannon. A cannon which fired a solid round shot of twelve pounds was described as a " Twelve Pounder," and so on for other sizes. Similarly smooth-bore muskets and guns were known by the weights of the bullets which they fired. Such bullets were of lead and spherical in shape. Their weights were sometimes given in ounces, but more often by the number of bullets which went to the pound. For example, a 2-ounce spherical bullet of solid lead was a No. 8, since eight such bullets made up a pound; a 1-ounce spherical lead bullet was a No. 16, because sixteen such bullets weighed a pound; and so on.

Nowadays a No. 8, or a No. 16 would be termed an 8-gauge (or 8-bore), or a 16-gauge (or 16-bore).

In Great Britain the size of the gauge or bore is actually measured at a point 9 inches from the breech, as laid down by the Proof Act and Rules.

In rifled arms the size is denoted by the actual diameter of the bore *across the lands*, that is by the smallest possible diameter of the bore (see Fig. 1).

In Great Britain and America this measurement (known as the **calibre**) is given in decimals of an inch, e.g. ·250, ·303, ·455, etc. But on the Continent it is given in millimetres, e.g. 6·35, 6·5, 7·9, 8, 9·3, etc.

There are two exceptions to both these general systems of measurement.

Smooth-bores of a smaller gauge than No. 32 are denoted by the diameter in decimals of an inch, e.g. ·410, ·360; and this applies generally if the bore is less than 0·5 inch.

Rifled Arms of larger calibre than ·600 are usually

classified according to their gauges, e.g. 10-bore, 8-bore, etc.

Parts of the Barrel.—The inside of the barrel of any arm consists of three parts.

The first of these is the **chamber** which is the portion at the breech end which accommodates the cartridge, and which is made to a shape which fits the particular type of cartridge which is used in the individual arm.

The chamber is almost always of considerably larger diameter than the actual bore, and is consequently connected to the bore by a taper.

In rifled arms this taper is known as the **Lead,** or **Leed.** The former spelling is more pedantically correct, as it denotes that portion of the inside of the barrel which *leads* from the chamber to the bore. But since the metal *lead* plays such a big part in shot and bullets, there is liable to be confusion of meaning if the same spelling is used for two such different things. For this reason the phonetic spelling, **Leed,** has come largely into use and will be adopted throughout this book.

The Leed is usually very short, that is less than half an inch in length.

In smooth-bores this connecting taper is called the **chamber cone.** In British guns the chamber cone is usually from $\frac{3}{8}$ to 1 inch in length, but in some Continental guns this cone may be as long as 2 inches.

The **bore** is that part of the inside of the barrel which lies between the front end of the Leed, or Chamber Cone, and the muzzle.

In a rifled arm this is the portion which is grooved, or rifled; and it is, or should be, perfectly cylindrical.

In a smooth-bore portions of the bore may be enlarged slightly after Proof in order to improve the shooting qualities of the weapon, and it is for this reason that the Proof Act fixes a point 9 inches from the breech for the purpose of measuring the gauge.

In the case of smooth-bore weapons with barrels shorter than 9 inches the gauge is, of course, as measured.

Breech Face.—The breech end of the chamber is sealed by a solid flat block of metal against which the barrel comes into position when the weapon is closed for firing, and which is termed the **breech face.**

SMOOTH-BORES

At the present time most smooth-bores are used only for firing charges of small shot, and by far the greater majority in existence are ordinary shotguns, or sporting game guns, used for shooting small game.

Shotguns are almost invariably made double-barrelled, the two barrels being placed side by side, although they are sometimes placed one over the other, when the gun is called an " Over and Under " gun.

Double-barrelled shotguns are made with outside hammers and with hammerless actions, in which the hammers are enclosed inside the body of the action. The latter are more costly and these are also made " Ejector " and " Non-ejector." The former automatically fling out (eject) the empty cartridge case after firing, and when the gun is opened for re-loading; but with the latter the fired case, or cases, must be removed by hand.

Single-barrelled shotguns are cheaper than double-barrelled and are usually made with external hammers, although there is no reason why one should not be made with a hammerless action if required. The cheapest type of single-barrelled shotgun is made with a bolt action similar to that employed in military rifles.

The outward appearance of ordinary double hammerless, and over and under hammerless guns are shown in Plate II.

Shotguns are also made in the form of single-barrelled magazine " Repeaters " with a " trombone action " slide under the barrel which must be moved back and then forwards in order to place a fresh cartridge in the chamber and close the breach. Such guns are popularly termed " Pump Guns."

" Automatic " shotguns are also made which are called " Auto-loaders " in America. The principle types are the Remington and Winchester in America; and the Browning, Sjörgen and Walther in Europe.

Smooth-bores are also made in the form of pistols, and such pistols are known as Shot Pistols. They are usually single barrelled, but are also made double barrelled. In general appearance they are similar to small shotguns, but with pistol stocks instead of the usual gun butts, or stocks.

There is a third type of smooth-bore which is almost a cross between a gun and a pistol, and this is a " Sawn-off shotgun," or a gun in which the barrels have been shortened to but a few inches (about 12) in length, so as to increase its portability. This type of arm is very popular with American gangsters.

The effective range of all shotguns and shot pistols is comparatively short, as the shot pellets which make up the charge disperse outwards after leaving the muzzle of the weapon, and owing to their light indi-

PLATE II

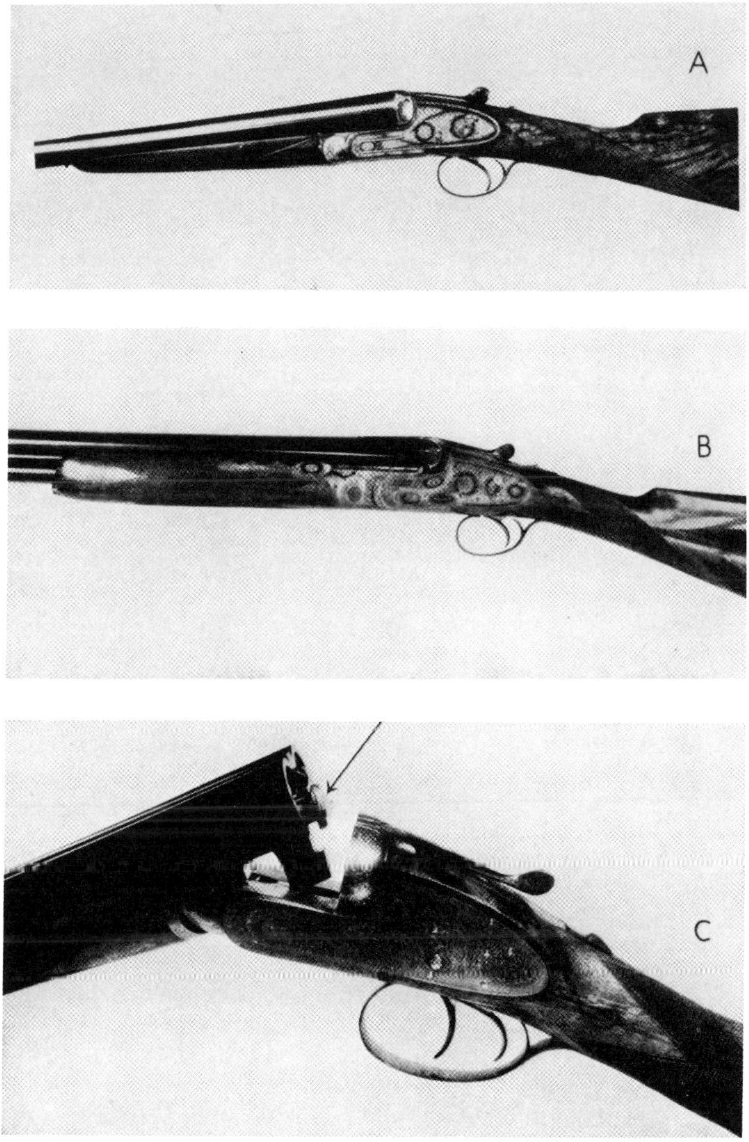

(A) A HIGH-GRADE DOUBLE-BARRELLED HAMMERLESS 12-BORE SPORTING GUN IN WHICH THE BARRELS ARE PLACED SIDE BY SIDE

(B) A HIGH-GRADE DOUBLE-BARRELLED HAMMERLESS 12-BORE SPORTING GUN IN WHICH THE BARRELS ARE PLACED ONE OVER THE OTHER

(C) A HIGH-GRADE DOUBLE-BARRELLED HAMMERLESS SPORTING GUN OPEN READY FOR LOADING

The arrow indicates the extractors

vidual weights they are unable to overcome air resistance, and so soon lose their velocity. But the very fact that the shot spreads renders marksmanship more easy than when a single projectile is used, and this fact, combined with the great possibility of injury conferred by the number of pellets, together render smooth-bores as peculiarly deadly weapons.

Single, solid and spherical lead bullets can be fired from almost any smooth bore; but the accuracy is very limited, while the difficulty of shooting is greatly increased. For these reasons such bullets are not likely to be used often by criminals, but the possibility of their use should not be forgotten.

Another class of smooth-bores comprises those weapons which are commonly known as "Saloon" rifles and pistols. They are almost invariably of about ·22 calibre and fire very short rim-fire cartridges which are termed "Bulleted Caps." The projectile is a bullet, but the charge consists of the cap priming. Such weapons have a comparatively short range and a very low degree of accuracy; but they are sufficiently powerful to kill human beings. "Saloon" pistols are sometimes sold as "Toy" pistols; but they are lethal weapons and cannot in any sense be regarded as toys suitable for children.

RIFLED ARMS

All rifled arms fire a single projectile, or bullet; while the purpose of the rifling, as is generally known, is to give the bullet a spin about its longitudinal axis during its passage along the bore. On leaving the muzzle of the weapon the bullet continues to spin, and this gyroscopic action maintains the bullet in stable nose-on flight during its passage through the air, which reduces air-resistance and adds enormously to the degree of accuracy obtainable, even at long ranges.

Rifled arms comprise Sporting Rifles, Military Rifles, Target Rifles, Miniature Rifles, Revolvers and Pistols.

Sporting Rifles.—Modern sporting rifles are made in all calibres from ·22 to ·600, and with numerous forms of action. But by far the greatest majority of sporting rifles intended for deer stalking or big-game shooting are made with bolt actions and magazines on similar lines to military rifles. Such rifles, however, are also made with double barrels, both hammer and hammerless, when their appearance and lines are very similar to shotguns. They are also made as so-called "automatics," and occasionally as single-shot single-barrelled weapons, that is they have no magazines but need to be reloaded after every shot. The actions then used are either that known as the Falling Block, or else the old Martini. But since neither of these actions is very suitable for powerful modern cartridges, they are now seldom encountered.

Much lighter, and less powerful, sporting rifles are also made for use against rabbits, rooks and such small animals. These weapons are invariably single barrelled, and are either fitted with hammer or hammerless actions like those of shotguns, or else with martini or bolt actions, while others have magazines and still others are so called "automatics." These last two types of "miniature" rifles are always made in ·22 calibre; but the former types are also made in ·250, ·275, ·295, ·300, ·310, ·320 and ·360 bores, although even in their case the ·22 is far the most common.

Military Rifles.—This group comprises the various Service Rifles of the different countries. All are made with magazines and bolt actions of various types, and the calibres range from ·256 (6·5 mm.) to ·311 (7·9 mm.).

Target Rifles.—Rifles used for full range target shooting, that is from 200 to 1,200 yards, or even more, are almost always specially selected military rifles of some sort or another, although the wooden butts and wooden portions adjoining the barrel are sometimes reduced or altered in form. But for all practical purposes these rifles may be regarded as good quality military rifles.

Miniature Rifles.—These rifles are generally of ·22 calibre and are used for Miniature, or "Small-Bore," target shooting. They are invariably single loading weapons, and are commonly made in Great Britain with Martini actions although in America some forms of bolt action are also popular. They are purposely made very heavy, comparatively speaking, so as to enable the shooter to hold his weapon steady with less effort, as too light a rifle fails to settle down in the hand and, contrary to popular belief, is much more difficult to hold absolutely still.

Revolvers.—We now come to the class of rifled arm which is intended to be used with one hand only. The first of this class is the Revolver, which is really a form of repeating pistol. It derives its name from the cylindrical magazine which rotates immediately in rear of the barrel. This "cylinder" usually holds five or six cartridges, each in a separate chamber, and each chamber can be aligned in turn with the bore. After firing a shot the hammer is cocked, and this cocking of the hammer rotates the cylinder until the next cartridge is in the proper position for firing. When this one has been fired the operation is repeated. The shooter is thus able to fire five or six shots with great rapidity and without reloading.

Revolvers are made with " single action " and " double action." In the single action type the hammer *must* be cocked by hand after each shot, and the weapon is then fired in the usual way by pressing the trigger. In the double action type the hammer *can* be cocked by hand, but it can also be cocked by a prolonged pull on the trigger. Thus with this type of revolver the full complement of shots can be fired merely by pulling the trigger after each shot.

Revolvers are also made with hammerless actions, in which the hammer is contained inside the actual action, and is hidden from view. Such revolvers must obviously be double-action weapons. The better-class hammerless revolvers are all fitted with some sort of safety device to prevent accidental discharge when the weapon is being carried loaded.

Revolvers are made in calibres which vary from ·22 to ·455.

Self-Loading Pistols.—The next type of pistol is the so-called " automatic." I have purposely used the words " so-called," as strictly true automatic pistols are only made by Mauser, and possibly one or two other makers. But such are really " sub-machine guns."

An automatic firearm is one which will *continue to fire, and go on firing, as long as the trigger is held back.* Machine guns are true automatics.

A weapon in which the discharge is utilised for ejecting the fired cartridge case, reloading a fresh live

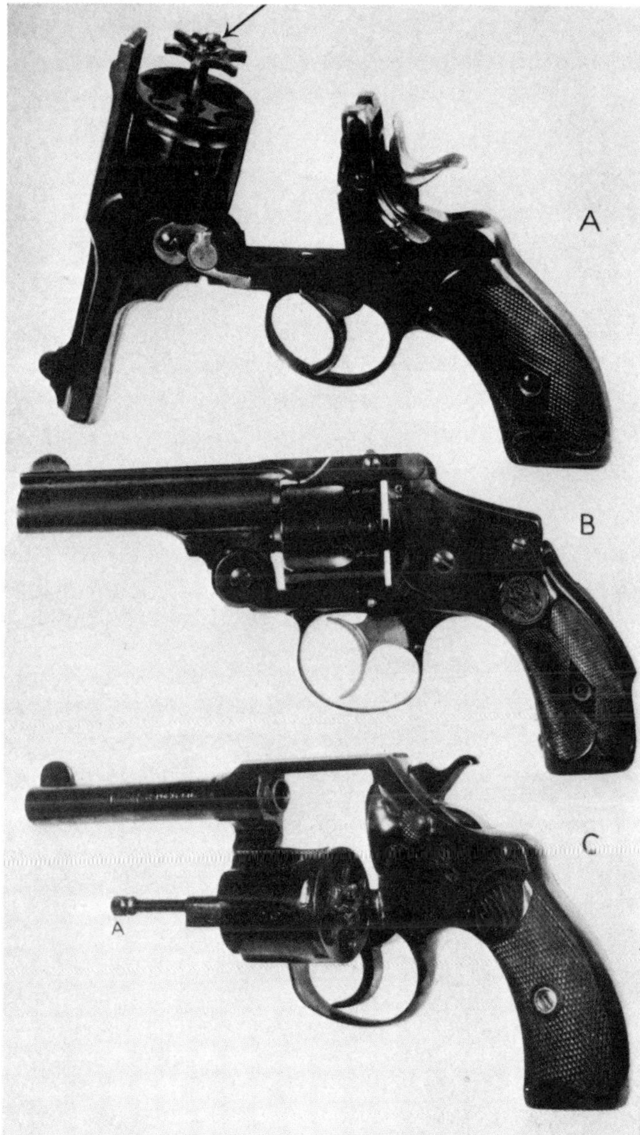

(A) A Webley "break down" model Revolver open after firing
The arrow indicates the extractor
(B) A Smith and Wesson hammerless "break down" model Revolver
(C) A Colt "solid frame" model Revolver open ready for loading
Extraction is effected by pushing to the rear the rod indicated by letter A

round in the chamber and cocking the action so that the weapon is automatically rendered ready to fire again, is termed a "Self-Loader." *In all self-loaders the trigger must be pressed anew for every shot that is fired.*

Almost all so-called automatic pistols, and rifles too, are really self-loading weapons. It may appear to verge on pedantry to differentiate thus, especially in view of the almost universal and popular use of the term "automatic." But it must be remembered that the difference between an automatic and a self-loader is really the difference between some form of machine gun and a pistol, so I do not think it unreasonable to advocate what is really nothing more than ordinary accuracy of nomenclature.

Self-loading pistols are totally different in appearance from revolvers; they have no cylinder, but carry their cartridges in a vertical magazine (which usually holds six to seven) in the stock; and are flatter than revolvers.

The makes and variations of self-loading pistols are legion, yet all makes bear a strong family likeness one to another. In calibre they range from the baby ·167 (4·25 mm.) Lilliput to the ·455 Webley and Scott.

Plate III shows the appearance of typical revolvers and Plate IV some typical self-loading pistols.

There is one other type of weapon which deserves mention as it is really the connecting link between self-loaders and automatics. This is the Sub-Machine Gun, the best known of which is probably the Thompson. This weapon is fitted with two pistol stocks and fires a ·45 self-loading pistol cartridge. The cartridges are contained in a helican magazine, which is made to take

100 at a time, and the weapon can be used *either* as a self-loader *or* a true automatic. It is a favourite weapon with American gangsters. Its overall length is but 2 feet and its weight 9 lb.

Single-Shot Pistols.—Pistols are also sometimes made as single-shot weapons, and such are generally either ·22 pistols or else shot pistols. The " Derringer " is really a single-shot pistol of a large calibre, usually ·41 or ·44.

EXTRACTION AND EJECTION

After a cartridge has been fired the empty case must obviously be removed from the chamber before the weapon can be reloaded with another round. In the majority of modern firearms the mere fact of opening the breech after firing automatically ejects the fired case; but in others the fired case is only partially unseated from its position in the chamber, and has to be removed by hand. This is the difference between ejection and extraction.

And since the effects of both extraction and ejection on fired cases can have a most important bearing on the work of identification of an individual weapon, a thorough understanding is essential of the broad principles involved in bringing these ends about.

These principles are really very simple.

In all " break-down " action guns and rifles, that is in all such weapons which open in the manner of an ordinary shotgun, extraction is effected by a portion of the breech end of the barrel being made separate from the rest of the barrel. On the arm being opened this portion is actuated by a cam and moves to the

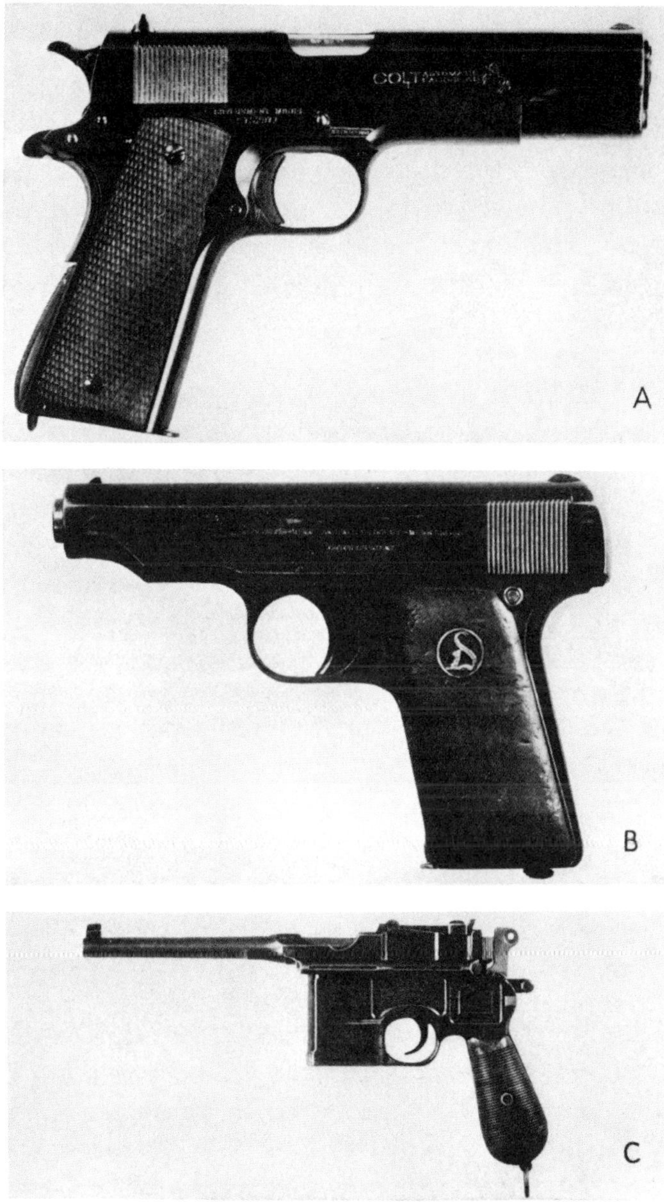

(A) A Colt ·45 hammer Self-loading Pistol

(B) An Ortgies ·32 hammerless Self-loading Pistol

(C) A Mauser 7·63 mm. Self-loading Pistol

rear. And since the head of the cartridge case is fitted with a rim, the moving limb of the barrel withdraws the cartridge with it. This movable limb is called the extractor.

Ejection of the fired case is brought about by a special lock mechanism which only comes into action when the hammer falls. This lock actuates the extractor and flicks it backwards at the moment when the arm is opened, and so ejects the fired case.

The extractors of an ordinary game gun can be seen in Plate II.

A similar extractor is used in all rifles with " falling block " or Martini actions, but in their case the extractor is actuated by the lever which opens the breech, and ejection is brought about either by the force with which this lever is jerked down, or else by a spring which is only freed from compression by the fall of the hammer.

In revolvers there is usually a common extractor for all the chambers in the cylinder, that portion belonging to each chamber being similar in general design to the extractor of an ordinary shotgun.

But in many of the earlier models of revolvers which are still in existence, as well as in some of the modern cheaper models, extraction is effected by rotating the cylinder so that each chamber is moved clear of the barrel in turn, when the fired case is pushed out from the front.

In all bolt-action rifles and shotguns extraction is effected simply by pulling the cartridge straight out. There is a small claw fitted on the circumference of the front end of the bolt, and this claw slips round the rim of the cartridge case. As the bolt is withdrawn it pulls the cartridge out with it.

Just before the bolt reaches the extreme end of its backward travel the front end passes over the metal support for the rear end of the magazine. The bottom part of the base of the cartridge butts up against this metal stop and the cartridge case is consequently pushed clear of the extractor claw. If the bolt is opened sharply this sudden butting of the cartridge base against the fixed metal stop is sufficient to fling the case clear of the rifle, and so bring about ejection.

In self-loading pistols the bolt is replaced by a sliding breech-block which is forced to slide backwards by the discharge in just the same manner as the bolt of a rifle is pulled back by hand.

The breech block is fitted with an extractor similar to that on a bolt-action rifle, and ejection is brought about by a small fixed metal block on the body of the pistol, the breech block being recessed so as to slide over it. This small metal block is called the Ejector.

In some makes of self-loading pistols there is no special ejector, and an existing part of the body is used to serve instead. Such parts of the body are the top edge of the magazine, or even the firing pin ; while in some makes the top cartridge in the magazine acts as the ejector.

The extractor of a Lee Enfield sporting rifle and the ejector in a self-loading pistol are also shown in Plate V.

CHAPTER II

CARTRIDGES AND THEIR COMPONENTS

MODERN cartridges differ greatly in outward appearance, but they are really all similar in that they consist of four main components: the cartridge case; the cap; the powder, or propellant; and the projectile, which may be either a single bullet or a charge of shot, when wads are also included.

Cartridge Cases.—The cartridge case, as its name infers, is the case which holds the other three components. Cases are made of brass, paper and brass, or copper, according to the type of weapon in which they are used. And since different weapons require cases of different shapes it is best to classify cartridge cases according to the types of weapons in which they are intended for use.

SHOTGUN CASES.—The great majority of shotgun cases consist of a brass head into which is fixed a stout paper tube which constitutes the walls of the case. The depth of the brass head is $\frac{5}{16}$ or $\frac{5}{8}$ of an inch in British cases, but in foreign cases it may be of any depth from $\frac{1}{4}$ to 1 inch. The principle, however, is the same in all.

In some cases there is a thin outer covering of brass which extends from the head almost to the top of the paper tube. Such cases are more costly and comparatively rare.

Some guns are designed to fire all-brass cases, in which there is no paper used at all. But the percentage of brass cases is small.

All shotgun cartridges have a rim of larger diameter round the extreme edge of the base. This rim fits into a groove at the extreme breech end of the barrel and so prevents the cartridge from dropping too far into the chamber and holds it in position close up to the breech face when the gun is closed.

The rim also enables the cartridge to be extracted or ejected easily and with certainty, as will be seen in a later chapter.

Rifle Cases.—The cartridge cases used in all sporting, military and target rifles except ·22 rifles are made of brass which is drawn out from the solid. In the majority of high powered rifles, including all military rifles, the mouth of the case, in which the bullet is seated, is necked down to a smaller diameter than the rest of the case. The result is a bottle-shaped appearance and increased accommodation for powder so that very high velocities can be obtained with small calibre bullets.

In other high-powered rifles the case is slightly tapered, but in all low-powered weapons the case is usually parallel, although it is sometimes bottle-shaped.

Rifle cases may either be " rimmed " at the base like shotgun cases, or " rimless." In the latter type there is no rim but just above the head there is a groove which runs round the circumference of the case. The cartridge is held in position in the chamber either by its taper, or else by the shoulder of the bottle neck, and the groove at the base is used for extraction as will be explained later.

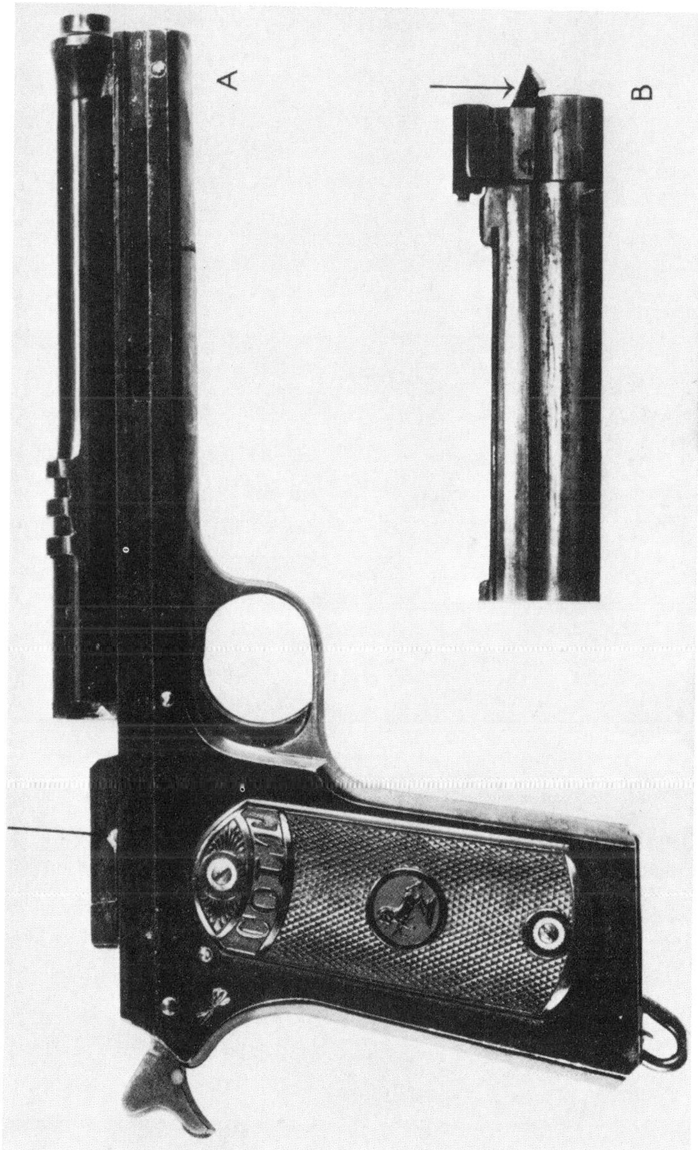

(A) A Colt ·38 Self-loading Pistol with the slide and breech mechanism removed
The arrow indicates the ejector block

(B) The Forward Half of the Bolt of a ·303 Sporting Rifle
The arrow indicates the extractor

The cartridge cases used in some special rifles built by Messrs. Holland & Holland combine the rimmed and rimless feature. The " rim " in this type of case is much deeper than the ordinary rim, and there is a groove at the extreme end as in ordinary rimless cases. Such cases are known as " belted rimless."

The cartridge cases used in ·22 " miniature " rifles are made of copper, straight and rimmed.

REVOLVER AND PISTOL CASES.—The cartridge cases used in revolvers and self-loading pistols are made of solid drawn brass and are termed " straight," or cylindrical, although they are really very slightly tapered. They are much shorter than rifle cases because they only have to accommodate much smaller powder charges.

Revolver cartridges are always rimmed, and modern self-loading pistol cases always rimless.

It should, however, be noted that in 1917 the famous American firms, Colt and Smith & Wesson, devised a special clip, which was made in two semi-circular sections, which allowed the use of ·45 Colt rimless self-loading pistol cartridges in revolvers adapted to take this clip. The object was to adapt revolvers to take the American service pistol cartridge. It is, therefore, just *possible* to fire a ·45 rimless pistol cartridge in a specially adapted revolver. This interchangeability also exists in the ·32 and ·25 calibres, as continental revolvers are now made to take these self-loading pistol cartridges.

The Mars 9 mm. self-loading pistol took a bottle-necked rimless case, and the Bergmann 6·5 mm. self-loading pistol a bottle-necked case without any rim or groove. But both these pistols are obsolete and ammunition for them is almost unobtainable. To the best of my belief these and the 8 mm. Schomberger, 7·65 mm. Bergmann, 7·65 mm. Luger-Parabellum, 7·63 mm. Mauser, 7·63 mm. Borchardt and 5 mm. Clement are the only exceptions to the general statement that self-loading pistol cartridge cases are straight.

Pistols and revolvers are also made to take the ·22 " miniature " cartridge, and there is no reason why a pistol should not be made to special order to take almost any low-power rifle cartridge.

Plate VI shows various types of shotgun, rifle, revolver and pistol cartridges.

CAPS

For all practical purposes the cap is an integral part of the cartridge case. In the vast majority of modern cartridges the cap is situated in the centre of the base of the case, and for this reason all such cartridges are termed " Central-Fire " cartridges. In ·22 " miniature " rifle cartridges, however, the cap is in fact a part of the cartridge case, the whole of the base being the " cap." These cartridges are fired by a blow on the edge of the base and on this account are known as " Rim Fire " cartridges.

In both central- and rim-fire cartridges the discharge is brought about by striking the cap either directly with the front end of the hammer, or else with the front end of a small rod, or " pin," which is knocked forward by the hammer, or by what corresponds to the hammer in bolt action weapons. The limb which actually hits the cap is called the **striker** or **firing**

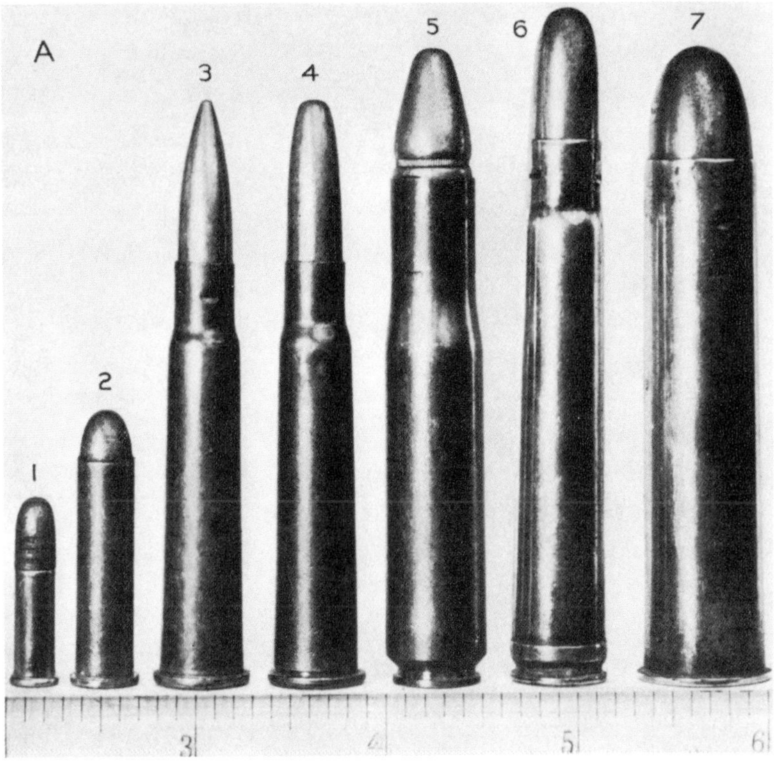

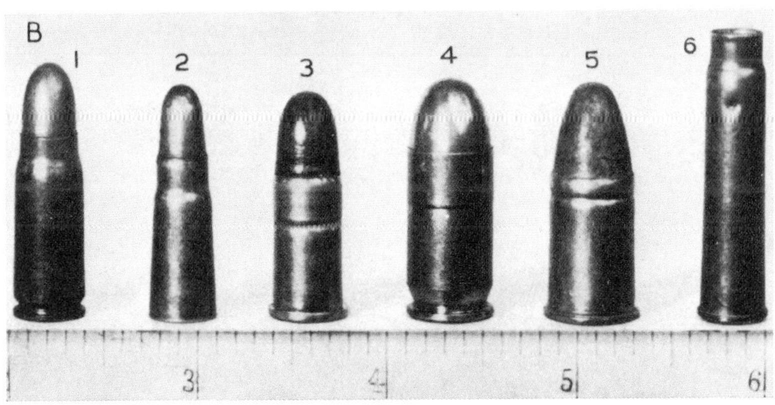

(A) Types of Rifle Cartridges

1, ·22 rim fire Long Rifle; 2, ·300 Rook Rifle, straight and rimmed case; 3, ·303 Mark VII, bottle-necked rimmed case and pointed bullet; 4, ·303 Mark VI, bottle-necked rimmed case and round-nosed bullet; 5, ·425 Sporting Rifle, bottle-necked rimless case and capped expanding bullet; 6, ·375 Magnum Sporting Rifle, bottle-necked belted rimless case and solid bullet; 7, ·577 Sporting Rifle, tapered rimmed case and solid bullet

(B) Types of Pistol and Revolver Cartridges

1, 7·63 mm. Mauser Self-loading Pistol, bottle-necked rimless case; 2, 6·5 mm. Bergmann Self-loading Pistol, bottle-necked tapered case without any sort of rim; 3, ·380 Revolver Cartridge, straight rimmed case; 4, ·45 Self-loading Pistol Cartridge, straight rimless case; 5, ·455 Service Revolver Cartridge, straight rimmed case; 6, 7·62 mm. Nagant (Russian) Revolver, bottle necked rimmed case with a nickel-jacketed bullet loaded with its nose level with the mouth of the case; the bullet is held in position by "stabbing," as can be seen

pin, and, as has just been explained, it may be either an integral part of the hammer or else a separate component.

There are a few other rim-fire cartridges besides the ·22, notably the ·41 and ·44 Derringer, but they are comparatively rare.

Before the central fire system was introduced the cap consisted of a brass pin which projected radially from the rim of the case. This pin fitted into a slot cut for it in the rear end of the chamber, and on firing the hammer fell vertically down on to the end of the pin and so fired the cap.

Weapons which fire these cartridges are called " Pin-Fire " weapons, and the cartridges " Pin-Fire " cartridges.

All percussion caps contain a small amount of explosive mixture, called the **cap composition** or **priming,** which is sufficiently sensitive to result in a chemical reaction being set up by the heat caused by a sudden blow. This chemical reaction changes the solid mixture into gases so rapidly that the change is, for all practical purposes, instantaneous; and consequently the gases form a very hot flame. It is this flame which ignites the powder charge just as a match lights a cigarette.

In central-fire cartridges the cap composition is contained in a small cylindrical capsule, closed at one end, which is placed in the centre of the base of the cartridge. This capsule may be either of copper, brass, or some other alloy.

POWDERS

Any firearm is really a machine for controlling the application of force which propels the bullet, or shot charge, through the air. The force necessary for this propulsion is generated by the very rapid production of gases resulting from the combustion of the powder charge; and on this account all powders which are used in firearms are termed **propellants.**

The subject of propellants is big and complex, and I will make no attempt to deal with it here beyond stating a few bare facts.

The propellants used in firearms are either Black Powder or Smokeless Powders. The latter are distinguished as Nitrocellulose Powders if nitroglycerine is not an ingredient; and Nitroglycerine Powders if nitroglycerine is an ingredient. All smokeless powders contain nitrocellulose.

Black powder consists of jet black, and rather shiny, grains.

Although black powder has been in use for nearly six centuries, and although improved methods of manufacture have naturally led to greater efficiency in action, its composition has remained practically the same in all countries.

In Great Britain this composition is:

Potassium Nitrate (Saltpetre) .	75 per cent.
Charcoal	15 ,, ,,
Sulphur	10 ,, ,,

The rate of burning is controlled by varying the size of grain, and the size of grain is denoted by a number.

The appearance of black powder is unmistakable to an experienced investigator, but if there is doubt a few grains can be placed on the tongue when the "salty" taste of the saltpetre will be noticed at once.

The real test, however, for black powder is to try to ignite a small heap of it with a match. It cannot be ignited with the flame, but will ignite immediately on being touched with the glowing end of a match left after the flame has been blown out. When black powder is used in a firearm it is unmistakable owing to the heavy smoke which is produced.

Powders based on nitroglycerine, nitrocellulose, or on a mixture of these two substances are almost entirely smokeless in action. They are all given the generic term of nitro powders and are legion in number.

All nitro powders used in rifles, pistols and revolvers are "gelatinised" powders, that is they are made by forming a "dough" into sticks or grains.

Cordite is the British service propellant and consists of nitroglycerine, guncotton and mineral jelly. It is of a brownish-yellow colour and in high-power rifle cartridges the charge consists of small sticks which are cut to a length to fit into the case. These "sticks" are really tubes in some cases, as a small hole will be noticed running up the middle of each. Such cordite is known as M.D.T. (modified and tubular).

Mark I cordite consists of solid cords, and in this the proportion of nitroglycerine is higher than in Cordite M.D. (modified) or in Cordite M.D.T., but the colour is very much the same, namely a brownish yellow.

"Pistol Cordite" consists of small cylindrical grains of cordite, about ·05 of an inch long, and is frequently used in service revolver cartridges and light rifles of the " rook and rabbit " class.

Moddite and Axite were two propellants made before the war which were both so similar to cordite in composition, behaviour and appearance that they could be regarded as cordite for all practical purposes.

Nitrocellulose powders are always granular. The grains sometimes consist of small cylinders, and sometimes of thin square flakes, when they are called " Flake " powders. They are frequently coated with graphite which gives them a dark grey colour. When not so coated the colour may be almost anything that the manufacturer fancies, but is usually a light grey.

Neonite is a British flake nitrocellulose powder which is now extensively used in revolver and pistol cartridges, and of a greenish grey colour.

Shotgun Powders.—Black Powder is seldom used in shotgun cartridges in Great Britain, although it is still loaded fairly extensively for native use abroad. All other shotgun powders are more or less smokeless and based either on nitroglycerine, nitrocellulose, or both, as are rifle and pistol powders. But they differ from these last in that they are also classified into " Bulk " and " Dense " powders.

The term " bulk " was applied to shotgun powders when black powder was replaced by smokeless powders. The standard charge for an ordinary 12-bore game gun was 3 drachms of black powder, and for convenience in loading powder measures were invariably used which held a volume of black powder which weighed exactly 3 drachms.

Smokeless powders were found to be much lighter in density than black powder, yet for purposes of general convenience in loading they were so standardised in manufacture that the same measure could be used for them as for black powder. In other words, the charge of smokeless powder was the same as that of black powder when measured by *volume*, or *bulk*, but not when measured by weight. The term " bulk " then came to be applied to all smokeless powders which could be loaded correctly with a black powder measure, and it has been in general use ever since.

Bulk powders are designated by the weights of charges which fill a 3-drachm black powder measure. For instance, there are 33-grain powders, 36-grain powders and 42-grain powders; and these designations mean that the volumes of the powders in question which exactly fill a 3-drachm black powder measure weigh respectively, 33, 36 and 42 grains.

Powders which cannot be loaded by bulk are termed " Dense " powders. Incidentally it may be noted that all rifle and pistol nitro powders are dense powders.

Bulk shotgun powders are also classified by the type of grain into Fibrous and Gelatinised powders, these latter being sub-divided into Porous and Non-Porous.

As a summing up the following definitions are given:

Nitroglycerine Powder.—A powder which contains nitroglycerine.

Nitrocellulose Powder.—A powder which contains nitrocellulose but no nitroglycerine.

Fibrous Powder.—A powder is called Fibrous when it is made from nitrocellulose and has a fibrous appearance when examined under a pocket lens. These fibrous grains can be broken down by rubbing in the palm of the hand.

Gelatinised Powder.—A powder which has been made

by forming a dough, or gelatinised mass, which is made into grains. The dough is made by treating a nitrocellulose mixture with solvents. This resulting dough-like mixture is pressed into tubes or rods, or rolled into sheets from which grains are cut.

Non-Porous Gelatinised Powder.—A powder made of grains which cannot be penetrated by the hot gases formed on combustion, and consequently such powders only burn on the surface. When viewed under a microscope the grains exhibit no "pores," or small holes in the surface.

Porous Gelatinised Powder.—A gelatinised powder which has been made porous during manufacture. When viewed under a microscope these pores can be clearly seen in the surface of the grains, although they can hardly be seen with an ordinary pocket lens.

Various types of powders are shown in Plate VII.

Powders, particularly shotgun powders, are so numerous that it is impossible to give a complete list of the different powders made in all countries. Some powders can be identified with comparative ease and certainty by visual examination ; but even for such identification considerable experience is necessary. Other powders are so similar, one to the other, that they can only be identified with certainty by means of a quantitive analysis, and there will not always be sufficient amount of the powder available for such an analysis to be made. In forensic work the actual identification of some particular powder will not often be necessary, but if such is necessary it is essential that a real expert in powders should be called in.

In British high-power rifle cartridges either some form or other of cordite is used, or else a granular

PLATE VII

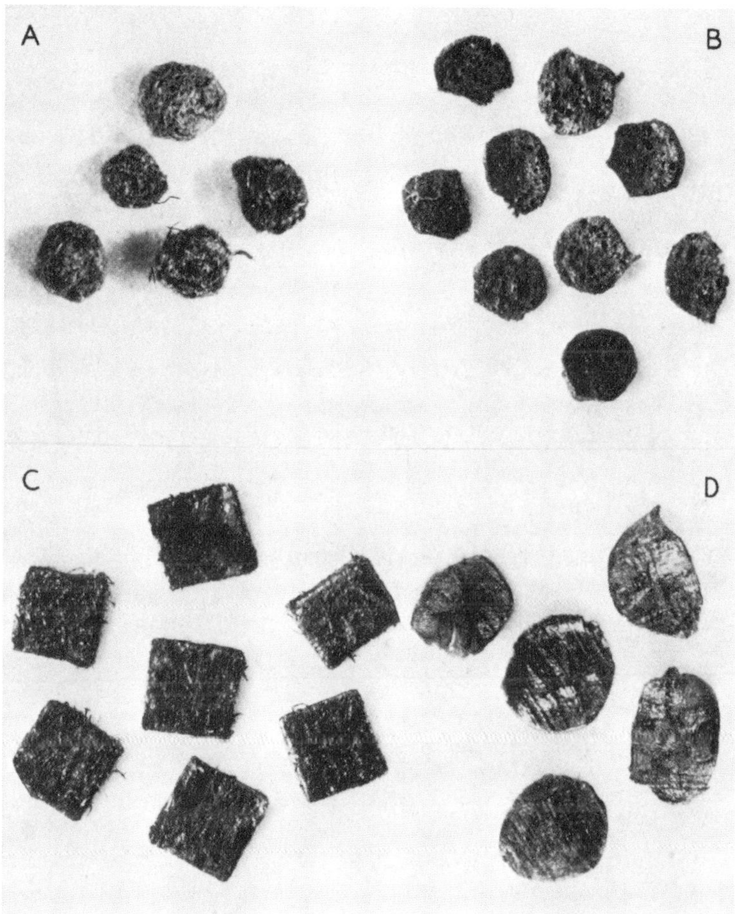

PHOTO-MICROGRAPHS OF SHOTGUN POWDERS

(A) E.C., a typical fibrous nitro-cellulose powder
(B) Smokeless Diamond, a typical semi-porous gelatinised nitro-cellulose powder
(C) Ballistite, a non-porous gelatinised nitro-glycerine powder
(D) Herco, another non-porous gelatinised nitro-glycerine powder

nitrocellulose powder comprising small cylindrical grains of a dark grey colour.

In British revolver and pistol cartridges either pistol cordite, or neonite.

The appearance of these powders has already been described.

Foreign high-power rifle and pistol and revolver cartridges are almost all loaded with some type of granular nitrocellulose powder.

The following are the shotgun powders which are in most common use, at any rate in Great Britain.

BRITISH SHOTGUN POWDERS

AMBERITE.—A 42-grain fibrous bulk nitrocellulose powder of a light grey colour.

BALLISTITE.—A dense gelatinised nitroglycerine powder. The grains are flat and square, and of a slightly shiny dark grey colour.

E.C.—A 33-grain fibrous nitrocellulose bulk powder. The colour is bright orange. When the grains are powdered in the palm of the hand there is a smell of camphor.

EMPIRE.—A 33-grain fibrous bulk nitrocellulose powder of a dull purple colour.

MODIFIED SMOKELESS DIAMOND.—A 36-grain porous gelatinised bulk nitrocellulose powder. The grains are *square* and the colour dull black.

NOBEL SMOKELESS.—A 33-grain fibrous bulk nitrocellulose powder of a bluish grey colour.

SCHULTZE.—An almost white 42-grain bulk fibrous nitrocellulose powder. The grains seem very hard when they are powdered in the palm of the hand.

SMOKELESS DIAMOND.—A 33-grain porous gelatinised bulk nitrocellulose powder. The grains are *circular* and slightly cupped. The colour is dull black.

AMERICAN SHOTGUN POWDERS

DUPONT OVAL.—A gelatinised non-porous dense nitrocellulose powder consisting of circular discs which are of a dark grey colour because they have been graphited.

DUPONT SMOKELESS.—A bulk partly gelatinised nitrocellulose powder consisting of round grains of a greyish white colour.

AMERICAN E.C.—A bulk fibrous nitrocellulose powder similar in kind and colour to British E.C.

HERCO.—A gelatinised non-porous dense powder containing a small proportion of nitroglycerine and consisting of circular discs which are of a dark grey colour because they are graphited. Each disc is marked with a cross.

BELGIAN SHOTGUN POWDERS

CLERMONITE.—A 33-grain fibrous bulk nitrocellulose powder. Somewhat similar in appearance to E.C., but the grains are smaller and of a more reddish colour.

COOPPAL NO. 1.—A 42-grain fibrous bulk nitrocellulose powder not unlike Schultze in appearance.

COOPPAL NO. 2.—A dense non-porous gelatinised nitrocellulose powder. The grains are square and flat, but much smaller than those of Ballistite although the colour is similar.

EMERALD.—A 33-grain fibrous bulk nitrocellulose powder of a bright green colour.

Mullerite.—A dense non-porous gelatinised nitrocellulose powder. The grains are square and flat and of a grey colour.

FRENCH SHOTGUN POWDERS

Poudre T.—A dense non-porous gelatinised nitrocellulose powder. The grains are square and flat, and the colour similar to that of Cooppal No. 2 and Ballistite, but in size they are intermediate between those two powders.

Poudre S.—A 42-grain fibrous bulk nitrocellulose powder very similar in appearance to Schultze or Cooppal No. 1.

GERMAN SHOTGUN POWDERS

Record.—A 33-grain porous gelatinised bulk nitrocellulose powder. The grains are similar to those of Smokeless Diamond, but less regular in size and shape and the colour is a deeper black.

Rottweil.—A dense slightly porous gelatinised nitrocellulose powder. The grains are flat and square and of about the same size as ballistite and of a similar colour.

Walsrode.—A dense slightly porous gelatinised nitrocellulose powder. The grains are round, slightly cupped and of a dark greenish colour.

SWEDISH SHOTGUN POWDER

Normal Flake.—A dense non-porous gelatinised nitrocellulose powder. The grains are square, flat and *very* thin. The colour is similar to that of Ballistite.

SHOT

In a shotgun cartridge the projectile consists principally of shot, although there are also some wads which ought to be included.

Shot, as is generally known, consists of small lead balls, or pellets. These pellets are of different sizes, and these sizes are denoted by numbers.

Shot is sometimes described as " Drop " or " Soft " shot and " Chilled " or " Hard " shot, but these names are really misleading. All shot, except the very largest sizes, is made by dropping molten lead from the top of a high tower into water ; and consequently all shot is both " Drop " and " Chilled " as the pellets are chilled by their passage through the air during their fall, and also by the water. The term " Chilled " was originally introduced to denote one particular make of shot in the manufacture of which blasts of cold air were projected on the pellets as they fell. This make of shot also consisted of a somewhat harder lead alloy than that usually used at the period, and consequently the name " Chilled " came to be considered as synonymous with hardness.

But nowadays there is little practical difference between the degrees of hardness of shot pellets made in different countries and by different firms. Experience in the field has shown the degree of hardness which gives the best results, and all manufacturers adopt this degree.

The lead used in shot is hardened by the addition of antimony, while a small percentage of arsenic is also added in order to make the molten lead alloy " run " to best advantage. If too little or too much arsenic is added the molten alloy will not form into round drops so readily, and consequently the percentage of arsenic is really fixed for manufacturers within narrow limits by practical exigency.

The very largest sizes of shot, commonly called "Buck Shot," are made by moulding and are consequently termed "Mould Shot."

In America and on the Continent some makers coat their shot pellets with a thin coating of either copper, or gilding metal (known in America as "Lubaloy"), which is an alloy of copper and zinc and has an appearance very similar to copper.

WADS.—It will be quite obvious that a charge consisting of a very large number of shot pellets must be restricted in some manner, or it would drop out of the cartridge. This restriction is effected by wads. In its simplest form wadding consists of a thick wad of some substance such as felt which is placed between the powder and shot charges, and a thin wad of cardboard which is placed on top of the shot charge and held in position by the end of the case being turned over against it.

The shot charge is thus held firmly in the cartridge case.

The thick felt wad, however, serves another essential purpose, namely it acts as a piston and seals the bore, thus preventing the expanding gases from escaping. This "obturation," or sealing of the bore, is of vital importance as on its efficiency depends the force which is exerted to propel the shot charge along the bore and through the air.

In order to lubricate the bore from round to round the thick felt wad is impregnated with grease, and since grease might affect the combustion of the powder, the latter is protected from the felt wad by a thin grease-proof card wad.

Similarly the shot pellets must be protected from the felt wad or some of them might be forced into the felt, and so another card wad is placed between the felt wad and the shot charge.

So it will be seen that the full list of components which are loaded into a shotgun cartridge case is the following:

(1) Powder Charge.
(2) Under-felt card wad. Usually $\frac{1}{12}$ inch in thickness.
(3) Felt wad. Usually $\frac{7}{16}$ or $\frac{1}{2}$ inch thick.
(4) Over-felt card wad. Usually $\frac{1}{16}$ or $\frac{1}{12}$ inch thick.
(5) Shot charge.
(6) Over-shot card wad. Usually $\frac{1}{16}$ inch, although $\frac{1}{20}$ and $\frac{1}{12}$ card wads are also used.

A section of a typical loaded British shotgun cartridge is shown in Plate VIII.

The chief variation from this system of wadding is to be found in the wads placed between the powder and shot charge. Various qualities of felt are used: white felt; dark brown felt; lighter brown felt. Sometimes cork is used in place of felt, and sometimes a material known as feltine, which is really a form of paper and of a slate grey colour.

Sometimes two thin felt wads are used instead of one thick one, and in many foreign cartridges the card wads are omitted altogether, their place being taken by thick paper discs glued to each side of the felt wad.

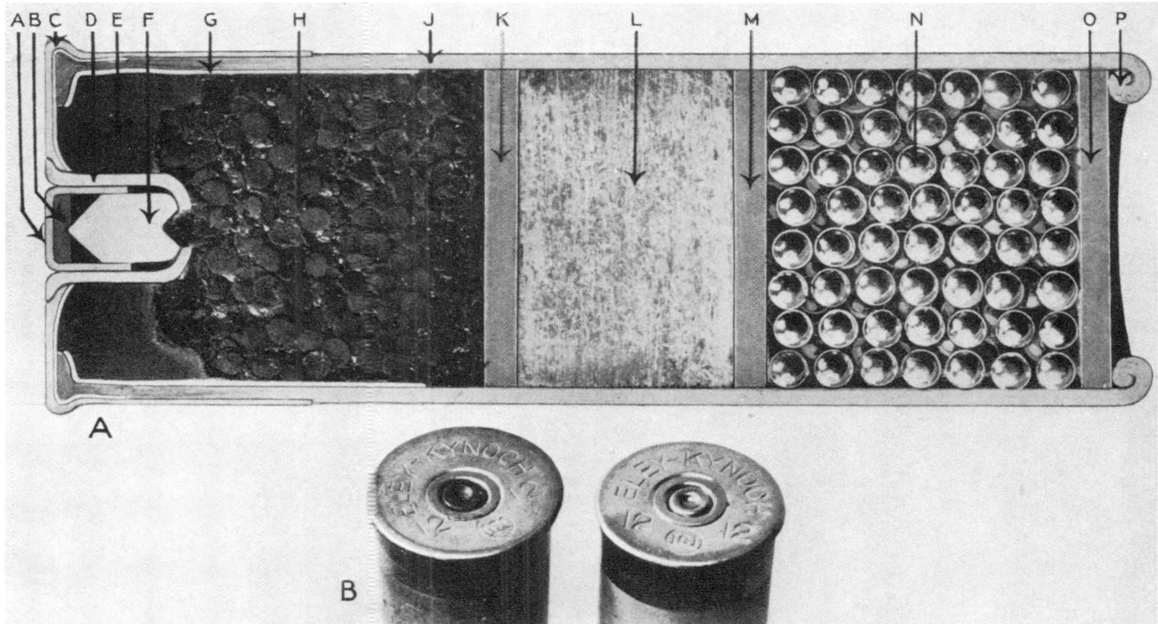

(A) A SECTION OF A TYPICAL BRITISH SHOTGUN CARTRIDGE

A, Cap Capsule; B, Cap Composition; C, Brass Head; D, Cap Chamber; E, Solid Rolled Paper Base; F, Cap Anvil; G, Metal Lining; H, Powder Charge; J, Paper Tube; K, Under-felt Card Wad; L, Felt Wad; M, Over-felt Card Wad; N, Shot Charge; O, Over-shot Card Wad; P, Turnover

(B) TWO FIRED 12-BORE SHOTGUN CARTRIDGES SHOWING THE EFFECTS OF VARYING PRESSURES ON THE CAP

The pressure developed in the case on the left was high, while that developed in the case on the right was normal

A recent substitution for the felt wad is called the "Hislop" wad and is really a cylinder of stout cardboard closed at both ends.

BULLETS

In rifles, revolvers and pistols (other than shot pistols) the projectile consists of a single bullet. In the days of black powder and before the introduction of nitro powders the bullets were all made of lead. And since it was found that pure lead was rather soft and exhibited a tendency to strip in the bore of the weapon, bullets were usually hardened with small additions of tin or antimony.

When nitro powders came into being, velocities were increased so much that leaden bullets could not grip the rifling in the bore, but stripped and were blown straight out. Consequently bullets were strengthened by the addition of an envelope, or jacket, of some harder and tougher substance. So it will be seen that the modern bullet consists of an outer envelope of some hard material and a core of lead, or lead alloy.

Various materials are used for bullet envelopes: cupro-nickel, steel with a thin cupro-nickel coating; brass with a thin cupro-nickel coating; and gilding metal. Envelopes made of this latter substance are known as "Nobeloy" envelopes in Great Britain and "Lubaloy" in America.

It should be noted that brass is an alloy of from 65 to 75 per cent. copper and from 35 to 25 per cent. zinc, and gilding metal an alloy of 85 to 95 per cent. copper and 15 to 5 per cent. zinc.

In Great Britain bullets envelopes are made of steel

coated with nobeloy; steel cupro-nickel coated; and cupro-nickel.

In America of lubaloy; cupro-nickel; and copper or brass, zinc coated.

On the Continent envelopes are made of cupro-nickel; steel, cupro-nickel coated; and brass, zinc coated. But this last is only used in some of the cheapest-quality self-loading pistol bullets.

The French Service Rifle bullet, Balle D, has no envelope and core, but is a solid copper alloy.

The British Service Rifle bullet Mark VII has a cupro-nickel jacket and a composite core, the point of the core being aluminium, and the main portion lead alloy. The object of the aluminium is to obtain a longer bullet, which gives increased accuracy, without adding to the weight. During the War, owing to the shortage of aluminium and the unprecedented demand, the aluminium tip was replaced by wood and jute.

In America the lead alloy bullets used in ·22 "miniature" rim-fire rifles are sometimes coated with lubaloy.

The bullets used in all high-power rifles are long in comparison to their diameter, the object being to obtain weight without offering a large surface to the resistance of the air, and so gain advantage at long range.

Revolvers and pistols are intended for close quarters, and the larger the effective diameter of the bullet the more efficient the stopping power. Consequently in these weapons the length of the bullet is short compared to its diameter.

The bullets used in revolvers are lead alloy, but those in self-loading pistols are jacketed as such bullets are less likely to become deformed and cause a jam during the automatic action of re-loading.

Many bullets have a circumferential groove near the base. The object of this is to allow the end of the case to be crimped into this groove, or " **cannelure,**" and so fix the bullet firmly in the case. Those bullets which have no cannelure are held in the case by " stabbing " the case at three points on its circumference, an operation which slightly dents the case into the envelope of the bullet at three points.

In lead bullets used in black powder rifles and other low-power arms there are frequently two, or even more, cannelures. But in their case these cannelures are filled with grease which lubricates the bore.

All bullets used in target and military rifles as well as in revolvers and pistols have solid noses, the lead core being fitted into the envelope from the base end, so as to conform with the Geneva convention against the use of expanding bullets.

But bullets used in sporting rifles are purposely made with various types of expanding points, or noses ; and such bullets are made by inserting the core into the envelope from the front end.

Military bullets are now always pointed so as to obtain enhanced velocities at long range, and some have slightly tapered, or " stream-lined," bases. The American, French and Swiss service bullets are all pointed and stream-lined. There is a British pointed and stream-lined bullet, but at present its use is confined to long-range Match Rifle shooting at Bisley.

When a granular powder is used in a rifle cartridge there is no wad. But in British sporting and military rifle cartridges, which are loaded with Mark I or M.D.T.

cordite, a thin cupped jute wad is inserted between the bullet and the powder which is held in position against the base of the bullet by the sticks of cordite. The object of this jute wad is to protect the base of the bullet and the leed and bore of the rifle from the blast of the very hot gases which are generated by cordite.

CHAPTER III

IGNITION, COMBUSTION AND PRESSURE

WE now come to a subject to which very little attention is normally paid, but a full and proper understanding of which is absolutely essential in forensic work.

As has already been explained in the preceding chapter the explosive charge in any cartridge consists of two separate parts: the Cap, and the Propellant, or Powder.

The cap contains a small amount of sensitive composition which can be exploded by the degree of heat resulting from friction, or a blow. This composition is contained in a cylindrical capsule of copper, or other fairly soft alloy, which is closed at one end; and it is this visible end which appears as the cap in the centre of the base of a cartridge.

When the cap is hit by the striker it is indented and the composition inside compressed between this indentation and a small piece of metal held in the open end of the capsule which is called the "Anvil." This sudden compression of the composition results in sufficient heat being generated to detonate the composition.

When this occurs the flame which is produced impinges on the powder charge in the body of the cartridge, and ignites it in exactly the same way as a match ignites a cigarette. So it will be realised that the sole, but none the less essential, function of the cap is to act as an igniter.

The powder charge is commonly said to "explode," but this is an incorrect description, for powder definitely *burns*; and this combustion, although rapid, is by no means instantaneous, and is slow compared to the detonation of the cap.

This process of combustion results in the solid powder being converted into gases rapidly, and at a high temperature; and it is the expanison of these gases which causes the explosive force which propels the bullet, or shot charge, along the barrel. This force is known as **pressure.**

Pressure acts in all directions. Not only does it propel the bullet, but also it forces the base of the cartridge case backwards against the breech face; expands the walls of the case so that they fit the chamber tightly; and acts outwards, submitting the barrel to stress.

In Great Britain pressure is measured in tons per square inch; in America in pounds per square inch; and on the Continent in kilogrammes per square centimetre.

The measurement of pressure is not difficult and in certain types of forensic investigations should be undertaken as a matter of course.

In order to measure the pressure developed by any particular cartridge a "pressure barrel" is used, which differs from an ordinary barrel chiefly in its greater thickness and strength. A hole is bored radially through the wall of the barrel at the point at which it is wished to measure the pressure. In this hole is inserted a metal plug, which has an enlarged flat head. This plug, or piston, is made of such a size that it is free to move in the hole, yet the fit is so close that when there is plenty of lubricating oil it forms a gastight plug exactly like the piston in the cylinder of an engine. The inner end of this piston is made flush with the surface of the bore.

Over the head of the piston on the outside of the barrel there is a fixed stirrup-like attachment which is made an integral part of the barrel, and through the top of this stirrup there is inserted a large screw with a flat lower end. This screw moves up and down radially to the bore in exactly the same way as the piston is free to move.

The diagram in Fig. 2 shows a longitudinal section through the middle of a pressure barrel with the piston and fixed stirrup.

The only other thing required beyond the actual cartridge is a "crusher." In the case of high-power rifles this is a cylinder of copper, and in that of low-power rifles, revolvers, pistols and shotguns it is of lead containing small amounts of antimony. It is placed standing on the flat head of the piston and the large screw which works in the fixed stirrup is screwed down on to the upper end of the crusher, which is thus held firmly in position as shown in Fig. 2.

When any pressure is generated inside the barrel the sides of the bore hold together, but the piston is free to move and is forced upwards by the expanding gases. As it rises the piston squeezes the crusher against the immovable screw, with the result that the crusher is compressed and changes its shape. The final length of the crusher is now measured with a micrometer, and the amount of reduction in length is proportional to the pressure which was generated. Tables are provided with every box of crushers which give a value for the pressure in tons for every thousandth of an inch reduction in the length of the crusher, so

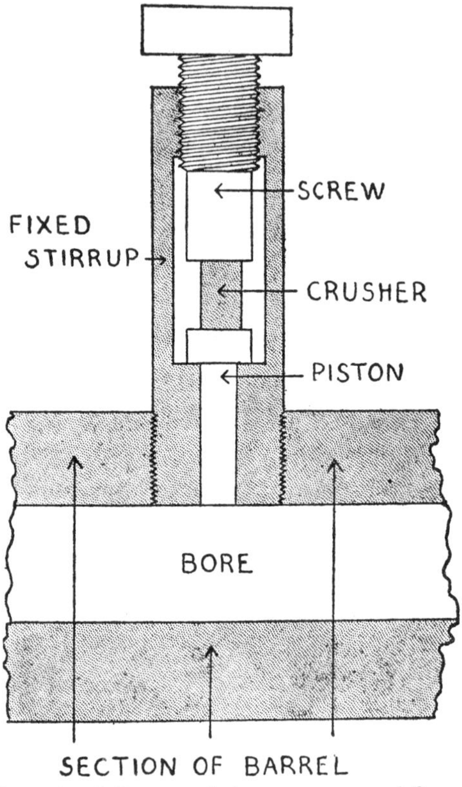

FIG. 2.—A sectional diagram of the apparatus used for measuring pressure.

that it is a simple matter to read off the pressure from the table directly the crusher has been measured.

This system measures radial pressures and is universally employed for shotguns, revolvers and pistols. But in the case of high-power rifles another system is also employed which measures the pressure generated on the base of the cartridge. The principle, however, is similar and a cylindrical copper crusher is used, on which rests a steel pad of greater diameter than the base of the cartridge. There is a longitudinal hole through the centre of this crusher to accommodate a special long firing pin, and the crusher is placed between the base pad and the cartridge and the breech face of the action. The pressure is obtained by measuring the reduction in length of the crusher as has already been described.

It will readily be understood that in order to obtain constant and comparable results it is essential that all possible sources of variation should be eliminated, and with this end in view everything connected with the measuring of pressures in shotguns in Great Britain was standardised in 1917, that is the diameters and weights of pistons, as well as diameters, lengths, weights and compositions of lead crushers and the distances from the breech face at which pressures were to be taken. This was a great advance, but the same standardisation was not brought about in the case of rifles, pistols and revolvers until 1924 when a most important conference was held after a lot of preliminary work, and effective standardisation was established.

In shotguns the maximum pressure occurs at about 1 inch from the breech face and so the pressure is always measured at this point, and when a particular batch of cartridges is quoted as developing such and such a mean pressure it is understood that this figure refers to the pressure at 1 inch from the breech face.

The same distance is adopted in rifles, but the cartridges in various low-power rifles are so short that the piston would not come between the base of the bullet and that of the cartridge unless it were placed nearer the breech face than 1 inch. In such rifles a shorter distance is, therefore, adopted.

Similarly with revolvers and pistols a distance of $\frac{3}{8}$ inch is adopted.

The cartridge is pierced with a small hole in such a way that this hole is exactly opposite the end of the piston when the cartridge is in the chamber.

There are other minor details such as the use of a gas check in addition to a piston when taking radial pressures in high-power rifles, and covering the hole bored in a brass cartridge case with adhesive paper, but such details are not of forensic importance. It is the *principles* that are important and which should be understood.

The measurement of the pressure developed by a few rounds selected at random from a batch of cartridges provides by far the best means of all for assessing the behaviour of those cartridges. A low pressure indicates sluggish combustion; a normal pressure, correct combustion; and a high pressure violence.

In shotguns there are many factors which affect the pressure which do not come within the scope of this book, but for ordinary practical purposes it may be taken that the most likely cause for variations in pressure is to be found in the cap.

It has already been explained that the cap is really an igniter. If the cap flame is weak, either because of a faulty cap or deterioration of the cap composition through long storage, the ignition of the powder will not be so effective as it should be and combustion will be more sluggish. This will be indicated by a low pressure. And if there is a tendency to faulty ignition this tendency may be increased by a weak striker blow. Consequently if for some reason it is desired to make tests with certain cartridges with a weapon other than that actually used in a crime, it is absolutely essential that the protrusions of the two strikers beyond the breech face when in the fired position should be carefully compared for identity, as well as the actual force of the blow delivered. This last can be done in a variety of ways, but probably the simplest is to place a rod in the bore which fits snugly without being tight, hold the weapon vertical, and snap the lock against the lower end of the rod which will be propelled upwards by the force of the blow. The vertical upward movement of the rod can be measured without much difficulty, and a very fair assessment of the strength of the striker blow is thus obtained.

Pressure is also affected by the tightness, or looseness, of the fit between cartridge and chamber, for the chambers of all firearms are purposely bored with a certain amount of tolerance in order to permit the use of a slightly over-sized cartridge and generally render loading more easy. In actual practice, however, it is doubtful whether the variations in pressure due to this cause would ever be sufficiently great as to affect the results obtained. Nevertheless, a capital charge is such a serious indictment that I feel that no precaution should be omitted which would ensure absolute reliability of results, and on this account I am of the opinion that the chamber dimensions of any test weapon should be compared with those of the crime weapon if tests are to be made.

If such tests necessitate the use of ammunition other

than that actually connected with the crime it is absolutely essential that at the very least one or two rounds of the crime cartridges, and still better a larger numbers be tested for pressure. Some of the test cartridges should be tested similarly, and if the two lots of result, do not correspond other cartridges must be obtained for the test which develop pressures similar to those of the crime cartridges. If the pressures are low and variable the caps may be tested for sensitiveness.

This last test is carried out by dropping a weight on the cap. The weight drops on to a striker of similar shape to that in a gun, rifle or revolver, and the cap should explode when a certain weight is dropped from a particular height. A special apparatus is used in which the weight is released electrically, and the actual weights necessary and the heights of the drops depend on the type of cap to be tested. The details of such tests are best obtained from the manufacturers, but it may be mentioned that in the case of the standard British shotgun cap the cap should explode when a weight of 2 ounces is dropped from a height of 14 inches, and in the case of the caps used in most revolver and pistol cartridges the test is given by a weight of 4 ounces falling from a height of 10 inches.

Detailed tests such as I have described may seem unnecessary to a layman who probably imagines that all cartridges behave in a similar manner. But to the trained ballistician they are a matter of normal routine, while their omission in any criminal investigation can only be regarded as a very grave blot on the conduct of a prosecution.

It may be, however, that sufficient "crime" cartridges are not always available for proper tests for pressure and cap sensitiveness. In such circumstances a single round can advantageously be tested for pressure ; and another opened to examine this powder and the cap of this cartridge may with advantage be tested for sensitiveness. Further, it should be realised that a very fair estimation of pressure can be obtained by a competent investigator from an examination of a fired cartridge case.

The pressure generated inside a cartridge case not only expands the sides of the case, but also drives the base back against the breech face of the arm. This last action flattens out the base, and the degree of flattening provides a very fair means of estimating whether the pressure was normal, low or high. Naturally the investigator needs experience ; but after one has lived, so to speak, with a pressure gun for years it is extraordinary how one learns to tell at a glance whether the pressure in any particular cartridge case varied from the normal.

Fig. 3 gives the outlines of longitudinal sections through the base of four typical central-fire cartridges. Fig. 3 (*a*) is the section of an unfired cartridge in which the base of the case and cap are shown. Fig. 3 (*c*) is the section of a similar fired case which developed a normal pressure. It will be seen that the cap has been flattened to a considerable extent round the striker indentation. Fig. 3 (*d*) shows the effect of an abnormally high pressure. Here the flattening of the cap is much more pronounced and the metal has been piled up round the edge of the striker indentation. Fig. 3 (*b*) shows the effect of an abnormally low pressure. The flattening of the cap is slight and the striker indentation more rounded at the edges and less deep than with a standard pressure.

Of course the effects of what may be termed standard pressures will vary in different cartridges. For instance, there will obviously be far more flattening of the cap in a ·303 Service rifle cartridge, in which the pressure is about 19 tons, than in an ordinary 12-bore shotgun cartridge, in which the pressure is under 3 tons. So

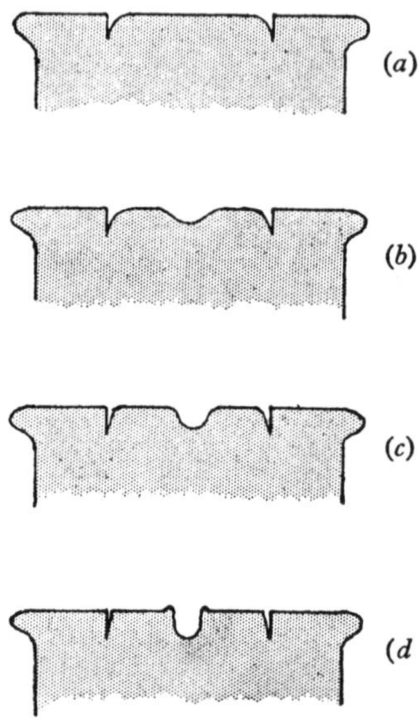

FIG. 3.—Vertical sections of cartridge cases showing the effect of pressure on the cap indentation.

(*a*) An unfired case. (*b*) A feeble pressure. (*c*) A normal pressure. (*d*) A high pressure.

comparisons can only be made between similar cartridges fired from similar weapons.

Another guide is to examine with a pocket lens the clearness with which the tool markings on the breech face are imprinted on the cap. Cartridges fired from the same weapon will frequently vary in this respect, and it should be quite plain that the clearer these markings are the higher the pressure which was developed.

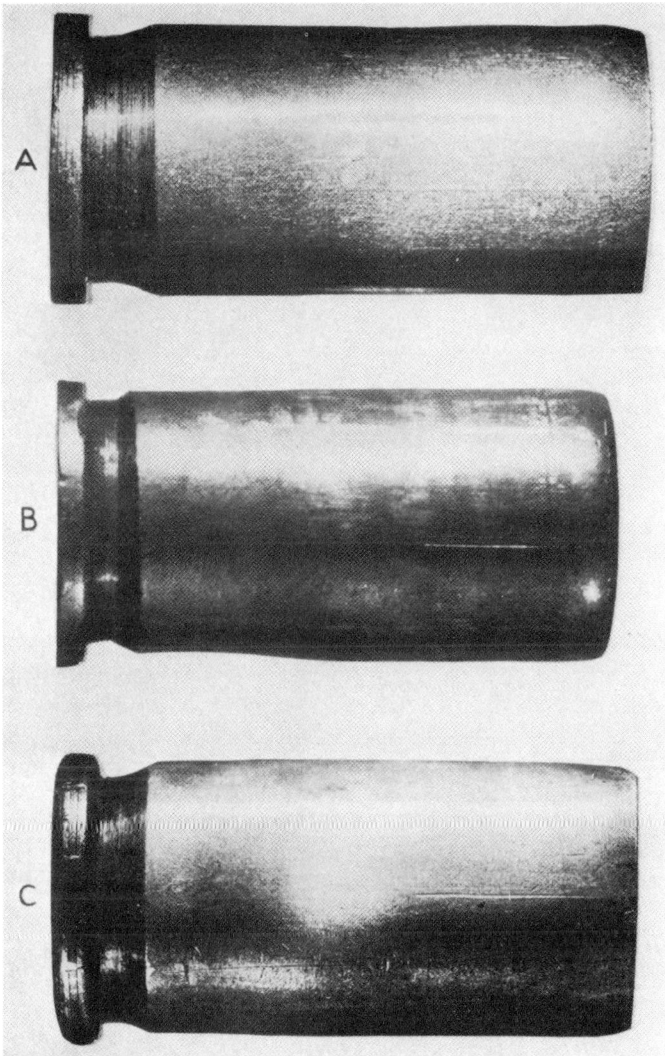

(A) AN UNFIRED 6·35 MM. CARTRIDGE CASE

(B) A 6·35 MM. CARTRIDGE CASE WHICH WAS FIRED IN A PISTOL WITH RATHER A LOOSE CHAMBER

Note the expansion of the forward part of the case. The longitudinal scratch should also be noted

(C) A 6·35 MM. CARTRIDGE WHICH WAS FIRED IN ANOTHER PISTOL

The chamber was tighter and there is less expansion of the case. But there is a longitudinal scratch very similar to that in the case shown in the middle photograph

Yet another guide to pressure is to hold the fired case up against the light and note the expansion of the head of the case immediately above the rim.

In revolver and pistol cartridges the walls of the case are much thicker at the base than half-way up. As a matter of fact this applies to all solid drawn brass cases, but in high-power rifles the pressures are so much higher than in revolvers and pistols that the effects of this difference in wall thickness is not so pronounced. In revolver and pistol cartridges, however, the front portion of the fired case is frequently expanded to fit the chamber tightly in the manner shown in Plate IX, and the distance to which this expansion of the case extends towards the base can offer a very good comparison for relative pressures in cases which have been fired from the same weapon.

In shotgun cartridges, and in some revolver cartridges and rifle cartridges fired in rifles with extractors of the shotgun type, there is usually the mark of the extractor imprinted round the head of the case. The distinctness with which this mark is imprinted is another guide to pressure.

Sometimes there is an escape of gas round the edge of the cap in a central-fire cartridge and the view is widely held that this is an indication of a high pressure. As a matter of fact it is far more likely to be the result of a *low* pressure, as will be appreciated when the cause is understood.

If a cap anvil which is very slightly on the large size is forced into a circular capsule the latter is deformed into a slightly elliptical shape. When this elliptical capsule is inserted into the head of the cartridge case it cannot fit the circular hole in the base of the case absolutely tightly, and on the combustion of the powder there is an escape of gas past the elliptical capsule.

But if the pressure is sufficiently high the capsule will be expanded to fit the hole in the head of the case quite tightly in spite of its original elliptical shape, and complete obturation, or sealing, will then be effected and the escape of gas prevented.

A gas escape past the cap does not necessarily mean that the pressure was appreciably below the normal; but it can be regarded as almost definite proof that the pressure was not violent, at any rate for the particular cartridge under consideration, for violence is a purely relative matter as has been seen, as a pressure which would be feeble in one cartridge would be violent in another.

Gas, however, may also escape through a split in the rim of the cartridge or through a split in the side close to the base. The former *may* be a sign of violence, and the latter generally is such a sign. If the breech face does not fit tightly against the breech end of the barrel, when the arm is said to be "off the face," the rim of the cartridge receives insufficient support and may easily be ruptured by a normal pressure. And so in the case of a split rim other indications of a high pressure must also be sought before forming any definite conclusion.

And now it may be asked quite legitimately: But of what practical use is all this from a purely forensic point of view?

I can but reply that since pressure is the most important of the ballistic elements it has the greatest influence on various seemingly unconnected matters, including the determination of the distance at which a fatal shot must have been fired and the identification of some individual arm by an examination of a fired cartridge case. This chapter deserves, therefore, the most careful study while the full importance of the subject will be better appreciated after completing the subsequent chapters.

CHAPTER IV

ASCERTAINING THE RANGE OF A FATAL SHOT

WHEN a human being is found shot there are four possible alternatives for the cause of the tragedy: accident, suicide, manslaughter, and murder. And before any very definite opinion can be formed certain problems must be considered in order to enable the authorities to formulate their conclusion. Probably one of the first of these problems will be the range of the fatal shot, that is the distance from which the shot was fired. It will, therefore, be but natural if we first turn our attention to this subject.

If the shot was fired from close quarters there will be definite evidence to this effect, such evidence being scorching of the skin or clothes immediately around the hole of entry; or "blackening" of this surface; or the presence on that surface of unburnt or partially burnt powder grains, or non-volatile products of combustion. Accordingly a most careful examination should be made for such evidence, preferably by a police officer or surgeon who is familiar with the appearance both of the various types of powder in normal use and the products of combustion which are likely to be in existence. This is an important point because the types of modern powders are so numerous and so utterly different one from another, that it might easily happen that someone without special training might fail to recognise powder as powder.

If the bullet has passed right through the body, and the direction of the wound is doubtful without a post-mortem, the surface round both holes should be examined.

If any evidence of the kind I have mentioned is found the range of the fatal shot can be determined with considerable accuracy, as will be appreciated from the following explanation.

Scorching.—This is caused by efflux of the very hot powder gases from the muzzle which are projected against the target. These gases are cooled very quickly on leaving the muzzle of the arm and so the range of their scorching effect is very limited. This scorching effect depends on :

(1) *Surface of the Target.*—That is whether the surface is wet, for some reason or other, or dry. My own experiments have been carried out with freshly flayed calf-skin from which the hair had been shaved, and ordinary cloth ; and I found that for all practical purposes it was impossible to detect any difference of results given by these two surfaces. But if clothes, or skin, were wetted, as by rain, the scorching range was reduced by about an inch.

(2) *Weight of Powder Charge.*—A big charge of powder will obviously produce a bigger volume of gas than a small charge, and so will have a greater effective scorching range. For this reason the scorching range of a high-power rifle is greater than that of a revolver or pistol.

(3) *Pressure.*—A normal, or high, pressure means that combustion is complete, which in its turn means that

a bigger volume of gas is developed. A low pressure will mean both that the volume of gas is smaller and that the actual temperature of the gases is lower, as the higher the actual gas pressure the higher the temperature of the gases in most cases.

(4) *Type of Powder.*—The gases developed by nitroglycerine powders are hotter than those developed by nitrocellulose powders.

There can be no possible doubt that all these factors have an influence on the scorching range, and in any case a proper understanding of the principles involved will help to the formation of a correct opinion. But for all practical purposes the extreme limit of the scorching range is so small that a reasonably approximate estimation of the range is good enough. With a service rifle scorching may occur up to 6 inches; and with a revolver or pistol up to 2 to 3 inches. So it can be assumed with absolute correctness that the presence of scorching is definite proof that the shot was fired from a range of but a very few inches.

Blackening.—This effect is the result of the deposit from the dirty powder gases, and is really very akin to scorching, the blackening range beginning where the scorching range ends. Blackening is not affected by a wet surface in quite the same way as scorching, although it can usually be removed very easily with a wet sponge, or rag. For this reason the absence of blackening round the wound in a body which had been in the rain for some time might be explained that the rain had washed it away, a point which should be remembered.

The blackening range is naturally affected by the weight of the powder charge in exactly the same way as the scorching range, that is the bigger the charge the greater the range. But the effects of pressure and the type of powder are different.

A low pressure means less complete combustion than a high pressure, and so the gases resulting from a low-pressure round will be more likely to make a dirty deposit than those from a high-pressure round.

Similarly the gases from the majority of nitrocellulose powders are likely to be somewhat dirtier than those from a nitroglycerine powder.

But as in scorching the range is really so close that for all practical purposes these niceties may be ignored, although it is as well that they should be understood.

Blackening with a high-power rifle, such as a service rifle, can occur up to about 9 inches; and with a revolver or pistol up to about 6 inches.

If the shot is fired with a cartridge loaded with black powder the blackening range is increased because of the smoke. On combustion black powder gives only about 44 per cent. of gaseous and 56 per cent. of volatile solid products, and it is this large percentage of volatile solid products of combustion which cause the heavy smoke. And the smoke is far more potent to deposit blackening than the gases generated by the combustion of nitro powders.

All self-loading pistol cartridges are smokeless, but revolver cartridges can be obtained loaded with black powder, and it should be remembered that it is possible by filing down the rim to fire a revolver cartridge in a self-loading pistol of the correct calibre. So the existence of smoke cannot be regarded as proof positive that a self-loading pistol was not used. But whether the cartridge were fired in a revolver or self-loading pistol the blackening range would be the same, and in the case of such a weapon is about 8 inches, while in the case of a shotgun firing 3 drachms of black powder the range is about 12 inches.

Unburnt Powder Grains.—Both scorching and blackening prove definitely that the shot was fired from very close quarters, in which case an assertion by the suspected person that the deceased fired the shot himself, cannot be disproved if the weapon used was a pistol or revolver. But if it is possible to establish that the range of the shot must have been greater than the length of the deceased's arm the matter assumes a somewhat different complexion, and the evidence may be of great use in bringing a murderer to book.

The extreme limit of the blackening range is well within any normal person's arm's length, and so the absence of blackening is no proof that the shot was fired from sufficiently far away to have made it impossible for the deceased to have been clutching either the weapon, or the individual who is suspected of having held the weapon.

However, the presence or absence of unburnt powder grains may indicate a range which is either just within or just without this critical distance; and on this account the investigation into the question of unburnt powder grains may become a matter of primary importance.

It has already been explained that the combustion of the powder charge is by no means an instantaneous process. This process results in the rapidly expanding gases carrying along with them, as they rush up the bore, some of that portion of the powder which has still to be consumed. It is this unburnt portion which can be expelled from the muzzle and projected forwards against any target which is sufficiently close. And since powder grains are heavier than either gas or smoke they attain a greater range, which means that they can be projected against a target beyond the extreme limit of the blackening range.

The proportion of the unburnt powder to the whole charge depends on three factors:

(1) The length of barrel of the weapon.

(2) The pressure.

(3) The type of powder.

These factors are important and deserve individual consideration.

LENGTH OF BARREL.—The longer the barrel the more time for the combustion to be completed, and consequently in a short-barrelled weapon there will always be a tendency towards unburnt powder being projected out of the muzzle, while in a long-barrelled arm this tendency will normally be so slight as to be almost non-existent. In fact, it may be assumed that with normal combustion there will be no unburnt powder grains at all projected by a rifle; and if any are projected from a shotgun it will indicate that the combustion was abnormal.

All rifle cartridges are loaded by the manufacturers at a factory where the importance of a perfect match between cap and powder is understood, while the work of effecting such a match has been brought to a fine art. For this reason combustion in a rifle will invariably be complete unless the cap has deteriorated through long storage under unsuitable conditions. But shotgun cartridges are frequently loaded by gunmakers, and although some gunmakers appreciate and understand the desirability of matching powder and cap there are others who do not understand such matters in the least. During the past fourteen years I have repeatedly come across shotgun cartridges which had been loaded by gunmakers in which combustion was hopelessly incomplete, and in which the fault lay in the use of cheap foreign cartridge cases capped with caps which were totally unsuited to the powders actually used. In the case of factory-loaded shotgun cartridges the match between powder and cap will be perfect. So in shotguns combustion is more likely to be incomplete than in rifles, and consequently it is possible that unburnt powder grains may be present round a wound from a shotgun. But when such powder grains are present it is proof positive that combustion was faulty, which will mean that pressure was abnormally low. And this fact should be remembered when conducting experiments to ascertain the dispersion of the shot charge.

I have strayed somewhat from the subject of the effect of barrel-length to that of combustion, but the two subjects are so interdependent that such temporary seeming irrelevancy was almost impossible.

To sum up, however, it can be stated that in normal circumstances the presence of unburnt powder grains will mean that the shot was fired from a revolver or pistol ; and that the shorter the barrel of the weapon used the greater will be the tendency to the presence of unburnt or partially burnt powder.

PRESSURE.—It has already been seen that pressure and combustion are interdependent : rapid initial combustion develops a high pressure which in its turn helps the combustion of the powder which is still unburnt by forcing the hot gases against and into the powder grains. Similarly sluggish initial combustion develops a low pressure which is unable to help the progress of combustion materially. In other words a lively pressure begets lively combustion, and a lively combustion begets a lively pressure; while the reverse is equally true.

So it will be realised that if two similar cartridges are fired from the same revolver and one develops a normal, or high, pressure while the other develops a low pressure, there will be a larger proportion of unburnt powder grains expelled in the case of the low-pressure round than in that of the high-pressure round.

TYPE OF POWDER.—As was seen in Chapter II powders differ in composition, size of grain and nature of grain. A fibrous powder, for instance, is much more easily ignited than a non-porous gelatinised powder; and consequently there will always be a greater tendency to incomplete combustion in the case of a powder of the latter type than in that of a powder of the former type.

Similarly the size of grain also has a marked effect—in fact, variation in the size of grain is the only method of controlling the rate of combustion of black powder, as has already been pointed out. This is really obvious, for a large volume of any burning substance will clearly take longer to be consumed than a small volume of the same substance. Hence the larger the size of powder grain in any cartridge the greater will be the tendency to unburnt grains, provided, that is to say, the same type of powder is used.

TESTING FOR UNBURNT POWDER GRAINS.—Let us now assume that someone has been shot and that the distance from which the fatal shot was fired is a vital matter for investigation. The revolver or pistol concerned will probably be available, as otherwise the presumption of murder will be almost inevitable. Let us also assume that the fired case and four unfired rounds are found in the revolver, or else that the fired case is picked up close by and four unfired rounds are found in the self-loading pistol, one in the chamber and three in the magazine.

No traces of any unburnt or partially burnt powder grains are found round the wound, and so tests must be conducted in order to ascertain the limit of the range at which unburnt powder grains are likely to be found round the wound.

If further investigations result in the finding of a box of cartridges similar to those found in the weapon the task is rendered simpler; but every precaution should be adopted to make sure that the cartridges in the box really *are* similar. The best plan is to consult the actual manufacturers of the cartridges if possible, who will be able to give the necessary information at once. The leading manufacturers in all countries imprint their name or distinguishing initials on the base of every cartridge sent out from their factories, and there are also minor markings which vary from year to year. Such markings will normally be known only to the manufacturers, as they are impressed for their own information; and by them it is possible for the manufacturer to say quite definitely whether two cartridges are of the same date or not.

If no other cartridges are found which seem to have a bearing on the case, a box of cartridges must be obtained of the same make, and if possible date, as those found in the weapon. If these latter cartridges are of British manufacture it is merely necessary to show one to the manufacturers in order to be provided with the necessary test cartridges. But if the cartridges found in the weapon are of foreign manufacture, and the makers have no agent in this country, no effort should be spared in procuring a box of cartridges in which the lettering and marks on the bases are identical both in form and position to those found on the cartridges taken from the suspect weapon.

After this the procedure is the same whether a supply of cartridges is found which appears to be connected with the tragedy, or whether a supply has been procured.

For the sake of convenience let us call the fired case of the round which caused the death the "crime" case, and the four cartridges found in the suspect weapon the "crime" cartridges; and the other supply of cartridges the "test" cartridges.

The first step is to examine the bore of the revolver to see whether there are any unburnt powder grains visible in it. This examination is best made with a small long-focus lens. It will probably be found that there are some unburnt grains present.

The next step is to examine the base of the "crime" case in order to obtain an opinion as to the pressure which was developed, that was whether it was high, normal or low for that particular type of cartridge.

Having done this one of the "crime" cartridges should be opened, the powder examined, identified if possible, and weighed, and the bullet weighed. If there seem to be an abnormally large proportion of unburnt grains present in the bore and the pressure in the "crime" case appears to have been low, the cap of the cartridge which has been opened can with advantage be tested for sensitiveness if the proper test can be carried out.

One of the remaining "crime" cartridges should then be tested for pressure, leaving two for the defence if the case comes up for trial.

If more "crime" cartridges are available, two, or as many as possible, can advantageously be tested for pressure; but a number should be left for the defence similar to that taken for test by the prosecution.

After this similar tests should be made with the "test" cartridges. That is one should be opened and the contents examined; one or more caps tested for sensitiveness if such a test has been made with a "crime" cartridge; and as many as possible tested for pressure. There will presumably be considerably more "test" cartridges available than "crime," and so the tests can be more complete.

A revolver should be procured of the same make, type, calibre and barrel-length as that found on the scene of the death and should be compared with the latter weapon for protrusion of the striker, gauge of chamber, strength of striker blow, and clearance between the front end of the cylinder and rear end of the barrel. There must always be a certain clearance here in a revolver through which an escape of powder gases is inevitable. In some cheap revolvers the degree of clearance varies considerably—I have known it vary by 100 per cent.—and this might influence the combustion differently in the two weapons.

Having obtained a weapon which pairs comparably with the "crime" weapon in the respects which I have mentioned the actual testing can be carried out. But the preliminary work is essential, more particularly the testing and checking for pressure. For it is an elementary principle, recognised universally by ballisticians in all countries, that the first step to take when making any comparative tests with two different lots of cartridges, is to test as many cartridges as possible taken haphazard from each batch by measuring the pressures actually developed.

The actual testing is really a more simple procedure than the essential preliminaries. All that has to be done is to fire shots with the "test" revolver and cartridges at a target of some material similar to the clothes of the deceased, or against white linen or freshly flayed bullock-skin with the hair shaved. This latter refinement is really unnecessary as all that is required is to ascertain the limit of range to which unburnt powder grains can be projected, and the nature of the surface used as a target cannot possibly affect this issue.

It is useful to fire the first shot with the muzzle almost touching the target so as to see the proportion of unburnt grains that are actually available for projection, and after this the range may be increased by three inches at a time. Information as to the scorching and blackening ranges may thus be obtained simultaneously.

The bases of all the fired "test" cases should be examined after every shot for indication of pressure; and if the pressure appears to be different from that which was probably developed in the "crime" case the round should be repeated.

By such means, and by such means only, is it possible

to obtain a scientifically accurate estimate of the "unburnt powder grain range." If any of the steps which I have described are omitted, entirely misleading results may be obtained. And in this connection it will be instructive to describe an actual demonstration which was so full of elementary mistakes that the result was as ludicrous as it was useless.

This demonstration was conducted not long ago under the ægis of the Director of Public Prosecutions with the apparent object of ascertaining the distance from which a fatal shot had been fired. The case was one of alleged murder and it was my lot to be present as an expert witness for the defence.

The weapon was a ·32 revolver and there was an ample supply of "crime" cartridges, although not enough to conduct a complete test for unburnt powder grains.

The defence was informed at the outset of the proceedings that the prosecution had taken the greatest trouble to obtain from the manufacturers a box of cartridges "from the very same *batch*" as those found in the possession of the accused. This in itself was startling, for although the manufacturers could always tell the *year* of any particular lot of cartridges, there is no possible means by which they could tell the *batch* unless the packing slip in the box was available, which it was not. One "crime" cartridge and one "test" cartridge were then opened and the bullets and powder charges weighed. I was handed the latter and saw at a glance that the two powders were quite different, both in colour and size of grain. I was then handed the two empty cases from which these different powder charges had been extracted, and saw what was obvious to anyone who possessed any familiarity with revolver cartridges, namely that the "crime" cartridge was of post-War manufacture while the "test" cartridge was of pre-War date! This could be told by the metal of the cap capsule. Before the War Eley cartridges had copper caps, while Kynoch had brass caps. After the War there was an amalgamation of these two firms and no copper-capped cartridges were made, although cartridges marked on the base "Eley" were made with brass caps.

No tests of any sort were made for pressure, nor were any comparative tests made between the "crime" and "test" revolvers.

But although these essential tests were ignored the greatest emphasis was laid on the fact that the "test" shots were being fired against pieces of material cut from the trousers of the dead man so as to ensure the material being identical with the coat which he had worn. As I have already pointed out this is a matter of no practical importance, but we were shown the tailor's name on the trouser buttons and coat in order to remove any doubt as to the exact identity of the material.

The first shot was very wisely fired at a range of a quarter of an inch and from this it was quite obvious that combustion was very incomplete, quite half the powder charge being expelled from the muzzle against the target in the form of unburnt powder grains. This fact clearly suggested faulty ignition, while a missfire a few rounds later provided corroborative evidence that the caps had failed.

Quite naturally unburnt powder grains were projected up to a surprising distance, for scarcely half the powder charge was ever being burnt, and when unburnt grains were definitely found at a range of 3 feet matters looked rather bad for the defence seeing that the accused declared that the shot had been fired accidentally in a struggle and that no sign of unburnt grains had been found on the dead man's coat.

But even if combustion in the " test " cartridges had been normal the demonstration was valueless because the powder grains in the " test " cartridges were *four times as large* as those in the " crime " cartridges. Both were " chopped " or " pistol " cordite, but of different brands, and the difference in size and colour of the grains could easily be seen from a distance of several feet.[1]

On leaving the demonstration the solicitor for the defence was very worried, but I assured him that the whole proceedings were farcical, and asked the manufacturers to date the samples I had been given from the " crime " and " test " cartridges. The " test " cartridges were identified as having been made in 1910 and the " crime " cartridges in 1921. And these were declared by the prosecution to be from the " same batch ! "

I made tests for pressure and found that the pressure in a " crime " cartridge was normal and 50 per cent. higher than that developed by two " test " cartridges! Also that a cap in a " crime " case passed the test for sensitiveness easily, while caps in " test " cases failed completely, proving that they had deteriorated with the years and that this was the cause of the incom-

[1] Samples of these two different powders from the " crime " and " test " cartridges, sealed at the time, are among my collection of " souvenirs."

plete combustion which I have mentioned as being so pronounced.

Later on I was informed unofficially that the prosecution had " been put wise " to their mistake. No reference to the demonstration was made either by counsel or in evidence, while it was stated by the prosecution that the fatal range was 3 inches. Naturally this statement was accepted by the defence without any cross-examination.

I was able to obtain some cartridges identical in every way to the " crime " cartridges, that is capped with the same caps and loaded with powder extracted from other cartridges of that date. These gave similar pressures to the " crime " cartridge, and were in fact identical. I found that at 12 inches unburnt powder grains were not present, and this shows what different results can be obtained by cartridges of different characteristics, while it emphasises again the vital importance of adopting the precautionary measures which I have explained.

Non-Volatile Products of Combustion.—These are projected in the same way as unburnt powder grains, but the factors governing their presence are quite different. For unburnt powder grains indicate incomplete combustion, and the less complete the combustion the more unburnt powder grains. But in the case of actual products of combustion the more complete the combustion the more products there will be, and consequently the maximum amount of these products can only be obtained when all the powder has been burnt up.

As a matter of fact all products of combustion are volatile and tend to cause blackening except one, barium nitrate. Metallic nitrates are frequently added to powders for reasons into which we need not enter here, but of the metallic nitrates which are so utilised barium nitrate is the only one which gives a non-volatile product of combustion. This product is barium carbonate which takes the form of very small, but hard, white globules. To the best of my belief the only powders which contain barium nitrate in any appreciable quantity are the shotgun powders. And so if these hard little white globules of barium carbonate are found on the surface surrounding a wound it is almost positive proof that a shotgun powder was used.

Spread of Shot in a Shotgun.—If the fatal shot was fired with a shotgun the problem of ascertaining the range from which it was fired is comparatively simple. This is because the pellets of a charge of shot do not fly through the air as a single cohesive projectile, but begin to disperse outwards very soon after their exit from the muzzle of the gun. This dispersion increases with the range, and consequently the greater the dispersion of the shot marks on any target the greater must have been the range from which the shot was fired against that target.

Now the degree to which the pellets of a shot charge spread out depends very largely on the individual gun used. The barrels of game guns are almost invariably bored with a slight constriction near the muzzle end, which is known as the "**choke**"; and the degree of constriction, or choke, has a marked influence on the subsequent spread of the shot charge. The amount of choke in game guns varies from none at all, when the gun is said to be bored "True Cylinder," to a constriction of $\frac{40}{1000}$ inch, when the gun is called "Full

Choke" ; and there are intermediate borings such as "Half Choke" and "Improved Cylinder," which is really the same as a very slight choke.

In order to assess the range of a fatal shot with real accuracy both the actual gun used and some cartridges similar to that used for the shot must be available. In such circumstances all that is necessary to do is to fire shots against a piece of leather stretched over a wooden frame at various distances until a hole is obtained in the leather of the same size as that of the entry hole of the wound. The range at which this shot was fired at the leather will fix that of the fatal shot.

If the gun is available but there are no "crime" cartridges, or else not sufficient to conduct a test such as I have described, certain precautions are necessary with which the reader will by now have become familiar. It must be realised that although the degree of choke in a gun exercises the chief control of the spread of the shot charge, this spread is also affected by the pressure developed in the cartridge. A high pressure tends to make the shot charge spread more rapidly, and a low pressure helps to keep it together. This being so, every precaution should be taken to ensure that the "test" cartridges develop the same pressure as the "crime" cartridge. The base of the fired case which was used for the fatal shot should be examined for indications of pressure, as should the bases of any other fired cartridges which may have been found on the scene of the crime. If other cartridges are found which were beyond doubt of the same lot as that used for the fatal shot, one should be opened to investigate the loading, and others should be tested for pressure. "Test" cartridges can then be obtained which will give identical pressures and such can be used with perfect confidence.

It may be urged, and with every reason, that the range of a fatal shot is probably so close that the degree of spread will be due entirely to the amount of choke in the gun, and that the influence of pressure will be almost negligible. And, on the whole, I agree with this view; and for this reason I do not regard the taking of the pressures in "test" shotgun cartridges as by any means so essential as in the case of "test" revolver or pistol cartridges, and I do not think that the omission of such tests could possibly be regarded as invalidating evidence. But at the same time there can really be no excuse for failing to include such tests, and their inclusion emphasises the part that the investigators were doing their best to eliminate every possible source of error. The testing of the pressures of a few rounds taken haphazard from every batch of cartridges loaded is, or certainly should be, a matter of normal routine for every competent loader of shotgun cartridges. And I cannot help feeling that a test which is demanded by ordinary sport should not be regarded as superfluous in any investigation when a human life may possibly be at stake.

If neither the gun nor any "crime" cartridges nor fired cases are available the range of the fatal shot can only be estimated. It may be assumed for all practical purposes that if the diameter of the wound is an inch, or less, then the distance of the shot was 18 inches or under, irrespective of the gauge of the shotgun or the degree of choke. Up to 2 feet there is very little difference in the spread between guns of various gauges and different chokes, the hole at this distance being slightly over an inch in diameter. At 3 feet the hole is nearly $1\frac{1}{2}$ inches in diameter, and the difference between the two extremes of boring, true cylinder and full choke, begins to be evident. At 6 feet the hole from a true cylinder would be nearly twice as big in diameter as that from a full choke.

The wads can also provide useful confirmatory evidence of the range, and for this reason the pathologist who makes the post-mortem cannot exercise too much care in trying to find every particle of wadding which was blown into the wound. On this account the pathologist should be familiar with the number and type of wads loaded in a shotgun cartridge, and this information should be checked if possible by unloading an unfired "crime" cartridge if any are available.

The over-shot card wad separates from the shot charge and rest of the wadding first of all, and seems invariably to have dropped clear of the shot charge before a range of 6 feet is reached. This being so, the presence of an over-shot card wad in a wound provides very strong presumptive evidence that the shot was fired from less than 6 feet; while the absence of the over-shot card wad suggests that the shot was fired from more than 6 feet provided the pathologist can be absolutely certain that he did not overlook the wad in the wound. This is a contingency which must be realised.

Effects of Striking Velocity of a Bullet.—There is one more method which is sometimes considered as a possible help towards the determination of the range of a fatal shot, namely the effect of the striking velocity of the bullet. A bullet travelling with a very high velocity will cause greater damage to tissue, and make a larger wound, than one travelling at a very low velocity. And since the velocity with which a bullet strikes any target depends on the range to that target, it has frequently been suggested that the nature of the wound gives an indication of the distance of the shot.

In the case of rifles it certainly is true that the range does have a big effect on the nature of the wound. But there can be no comparison between wounds caused by rifle bullets and those caused by pistol or revolver bullets. Modern military rifles have a man-killing range of over 4,000 yards and quite naturally there will be considerable difference in the nature of a wound inflicted by such a rifle at but half this extreme range and one inflicted at but 5 yards. But the muzzle velocities of such rifles are in the neighbourhood of 2,800 f.s. The British service rifle has not such an extreme range with the service cartridge, but its effective man-killing range is not far off 3,000 yards and its muzzle velocity is 2,450 f.s.

When the velocity of any bullet has dropped below that of sound (1,100 f.s.) its wounding effect is little changed by variations in velocity. The muzzle velocity developed by all revolvers is usually between 600 and 750 f.s., and at these velocities the changes in velocity due to range are comparatively slight at the ranges within which such weapons can be used with reasonable chances of hitting. The velocities developed by most self-loading pistols are similar, although there are a few such weapons which give higher velocities. But even with these the changes in velocity are slight within normal ranges. Consequently statements that a shot was fired from a distance of between 7 and 9 yards or between 10 and 12 yards are mere guesses, and ignorant guesses at that. Any medical man who made such a statement should be asked what was the muzzle velocity of the weapon used, and what the striking velocities at, say, 10, 15 and 20 yards. It would be a safe bet that he would have no notion of the answers to these questions, nor how to find them out. If he had he would never have made the original statement.

The truth is that once the "unburnt powder grain" range has been exceeded it is utterly impossible to say whether a shot was fired from 5 yards or 20, and that any opinion formed on the appearance of the wound can only be regarded as a pure guess.

It might be possible to differentiate between a shot fired at 5 yards and one fired at 100 yards in the case of a revolver, or pistol. But the definite statements as to the range which are sometimes made have no real scientific backing.

I have more than once been asked by members of the legal profession if it is not a fact that after a certain range the bullet suddenly begins to accelerate through the air. I can, therefore, only presume that there is some belief in this theory. So it can be stated quite definitely that such an idea is quite erroneous. A bullet leaves the muzzle with a certain velocity, but after its exit from the muzzle it loses velocity throughout its flight. The rate of loss in velocity becomes less rapid as the range is increased because the air resistance is greatest when the velocity is highest. But the bullet is losing velocity all the time.

It may, perhaps, make this more clear if I give the actual velocities at different ranges for a few typical pistols and revolvers, so I am including these values in the following table.

MUZZLE AND STRIKING VELOCITIES AT DIFFERENT RANGES FOR DIFFERENT PISTOLS AND REVOLVERS

Weapon and Cartridge.	Velocity in f.s.					
	Muzzle.	5 yds.	10 yds.	15 yds.	20 yds.	25 yds.
·455 Service Revolver	600	596·5	593	590	587	584
·45 Colt Self-loader 230 grs. bullet . .	800	795	790	785	780	776
·38 S. & W. Revolver.	625	621	617	613	609	605
·380 Short Self-loader or 9 mm. Browning (short)	850	842	834	826	819	813
9 mm. Parabellum .	1,200	1,185	1,170	1,155	1,140	1,135
·32 S. & W. Revolver	600	595	590	585	580	576
·32 Self-loader or 7·65 Browning . . .	875	868	861	854	847	840
7·63 mm. Mauser. .	1,400	1,378	1,356	1,335	1,315	1,296
·250 Self-loader . .	625	619	613	608	602	597

The values which I have given for the respective muzzle velocities are those developed by cartridges of British manufacture in revolvers or pistols of normal barrel lengths. Different lengths of barrel will cause slightly different velocities, an increase in barrel length causing an increase in velocity while a short barrel causes a loss. But various foreign manufacturers load their cartridges to somewhat different ballistics, and consequently there is ample scope for variations in initial velocity which are as great, if not greater, than the loss in velocity due to air resistance between ranges of 5 and 25 yards. Further, the velocities developed by individual rounds of any batch of cartridges may, and frequently do, show a bigger difference than the actual loss in velocity between 5 and 25 yards. This must merely emphasise the impossibility of a pathologist being able to ascertain the range of a fatal shot with any degree of scientific accuracy by deductions based on an assumed striking velocity.

It may be urged that the values for the various striking velocities which I have given have been calculated, and so are merely theoretical. This is true up to a point. But it must be remembered that the ballistic tables with which such calculations are made have been derived from great numbers of long series of shots, the results of which have been measured instrumentally. The curves for air resistance at different velocities are, therefore, as much empirical as theoretical; while the fact that a knowledge of the instrumental muzzle velocity of a rifle and its true zero elevation enables the ballistician to calculate the elevation required for 1,000 yards with sufficient exactitude to enable the marksman to hit the bull's-eye at this range with his first shot, cannot but indicate that the principles on which the ballistician works are not very far wrong.

Probably one of the most remarkable ballistic calculations which have ever been made was that which enabled the German artillerymen to shell Paris with "Big Bertha" without the necessity for any ranging rounds.

I do not, therefore, think it unreasonable to assume that the science of ballistics gives us the relative values of the striking velocities at different ranges for pistol and revolver bullets with absolute accuracy.

CHAPTER V

THE TIME OF THE FATAL SHOT AND OTHER PROBLEMS

A problem with which those investigating a crime are frequently confronted is to determine the time which has probably elapsed since the firing of a shot. A weapon may be discovered which may, or may not, have been used by the criminal. The barrel is clearly "fouled" and so the question arises, is it possible by an examination of the barrel to ascertain with any degree of accuracy the period of time which has elapsed since that weapon was fired?

Before making any attempt to answer this question it will be better if we consider first all the factors which are involved.

When any firearm is discharged the barrel is fouled both by the passage of the projectile along the bore and by certain products of combustion being deposited on the surface of the bore. If the bore is not cleaned the surface will become corroded with the lapse of time. This corrosion is really rust. Now rusting is not quite such a simple matter as is popularly supposed in spite of the apparent ease with which it occurs when least desired. But there are certain facts connected with the formation of rust which are beyond dispute. These are that moisture *must* be present; that the more moisture present the greater the formation of rust; that the formation of rust is helped by the presence of an acid; and that rusting is retarded by the presence of an alkali.

When the bore of any firearm rusts the necessary moisture is provided by the air, which varies in degree of humidity in different places and in the same place with different weather.

The products of combustion, however, can hasten or retard the effects of the humidity of the air. For example, the most potent factor in hastening rusting is one of the products of combustion of the *cap composition*, and not of the powder, as is popularly supposed. The majority of caps are primed with a composition which contains potassium chlorate. On the cap being fired this substance gives up its oxygen very readily and becomes potassium chloride, and this potassium chloride is deposited in the bore.

Now potassium chloride is a salt very similar to ordinary table salt or sea salt (sodium chloride) and like common salt has an affinity for water. Everyone knows how iron and steel will rust much more readily at the seaside, the reason being the presence of the salt which attracts the moisture from the air on to any surface on which it is deposited. In exactly the same way the potassium chloride in the bore attracts moisture from the air, and thus greatly helps the formation of rust in the bore.

But the proportion of potassium chlorate in the cap composition varies in different cartridges, and in some caps it is entirely omitted. Consequently the amount of potassium chloride deposited in the bore will depend on the type of cartridge used.

Similarly the compositions of powders differ from

each other and consequently their products of combustion. In many powders these products are neutral, that is neither acid nor alkaline. In a few powders they may be acid, while in black powder and many modern powders, including all British smokeless shotgun powders, they are alkaline with a view to checking the rusting effect of the cap composition.

Then, as has already been stated, the humidity of the air at any place is constantly varying, which means that the amount of moisture available to help the formation of rust is constantly varying.

So it will be appreciated that the formation of rust in the bore of a discharged firearm depends on:

(1) The type of cap in the cartridge.

(2) The type of powder.

(3) The humidity of the *surrounding* atmosphere since the time of firing until that of the investigation.

The first of these factors depends on a knowledge of the composition of the cap priming which can only be obtained, in many instances, by an analysis of the priming extracted from a similar cartridge to that used. The powder must also be identified. And even then there is the question of the humidity of the atmosphere immediately surrounding the weapon during its period of rest.

I have heard of claims being made that it is possible to fix the exact date of firing after a period of months even without knowledge of the cartridges which were used! If any investigator ever made such a statement in the witness-box he would merely proclaim himself as being completely ignorant even of the principles involved.

The truth is that it is utterly impossible to fix the

date of discharge with any scientific accuracy after the lapse of a comparatively few hours. It may sound very impressive to declare that a weapon has been "recently" discharged. But "recently" is too vague a word to merit very serious attention. A really experienced expert will probably be able to form a fairly accurate idea as to whether the amount of corrosion in the bore of a fired weapon indicated a lapse of days or months, but that is all. Of course in some cases a barrel may be so filled with rust that many months, or even years, must have elapsed since it had been used: but even then the opinion delivered can never be more than an intelligent estimate.

It *might* be possible to back this opinion by a chemical analysis of the rust, for rust usually contains ferrous oxide, ferric oxide, and iron carbonate; and if the rust has been long exposed to the air, the amount of ferric oxide and iron carbonate is relatively large, and the amounts of ferrous oxide small.

Yet another factor affecting the rusting is the type of steel used for the barrel, as some steels will rust more readily than others; while the presence of chromium in sufficient quantity, for instance, will prevent rusting altogether. In actual practice, however, the difference in the compositions of the steel used in the barrels of ordinary firearms is too slight to make any real difference; and I have merely mentioned the point in order to show how complex the whole question can be.

And a fired cartridge provides even less useful evidence in this connection, for brass does not corrode to nearly the same extent as iron or steel. Sometimes there is an escape of gas past the cap of a central-fire cartridge, and when this occurs the base of the fired case is blackened immediately round the point of the gas escape, and in time develops a verdigris coloration. The contrast between such blackening and the surface of the surrounding brass *might* give a very experienced expert some hint as to whether the cartridge had been fired within twenty-four hours or no; but here again his opinion could not be regarded as anything more than an estimate.

The truth is that it is absolutely impossible to ascertain with any scientific accuracy the period of time which has elapsed since a weapon or cartridge case has been fired.

ACCIDENTAL DISCHARGE

One of the commonest causes of shooting fatalities is the accidental discharge of a weapon which was loaded and at full cock, and not a single shooting season passes without some such fatal accidents being reported. Anyone carrying a gun may stumble and drop his gun, when the blow which the latter receives on falling is liable to jar the lock or locks off, and thus cause an accidental discharge. The possibility of this happening is almost too well known to need emphasis, but it may not be out of place to explain very briefly the general *principle* on which all gun, rifle and pistol locks work, so that the reader will be able better to understand *why* an accidental discharge is possible.

In reality the essential limbs of a lock work on the principle of a cog-wheel and ratchet, as may be seen in Fig. 4. It should be clearly understood that these diagrams are not intended to represent any particular type of lock, but merely the principle on which all locks function.

The two essential limbs consist of a wheel, or "Tumbler," A, which is fixed to an axle, B, which in turn rotates in two bearings in the lock. This wheel is

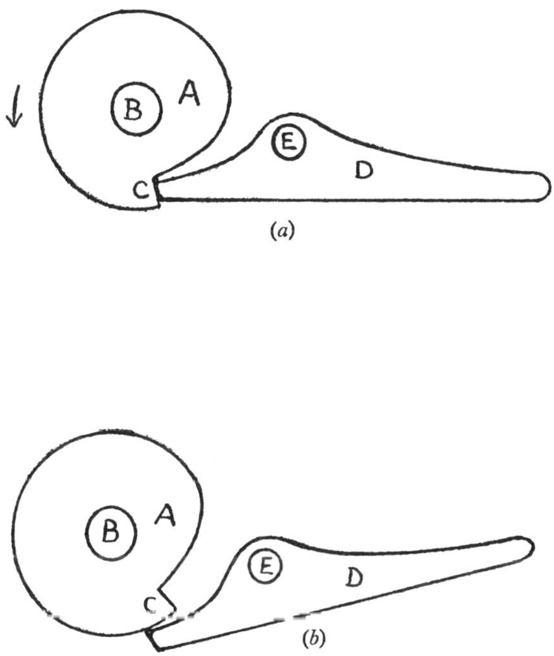

FIG. 4.—Diagrams to illustrate the principle on which a shotgun lock works. A, Tumbler. B, Tumbler axle. C, Bent. D, Sear. E, Sear peg. (*a*) Cocked position. (*b*) Fired position.

actuated by a spring, termed the "Main Spring" (not shown for the sake of simplicity) which is compressed when the lock is cocked and then tends to rotate the tumbler and axle in the direction of the arrow. There is a cog, or "Bent," C, in the lower edge of the tumbler; and in this bent there fits a ratchet, or "Sear," D, which pivots about a peg, E. The front end of the sear which actually fits in the bent just as a ratchet does in a cog, is termed the "Nose," while the other end of the sear is called the "Tail." The sear is also actuated by a spring, known as the "Sear Spring," which depresses the tail and so tends to keep the nose in the bent, as shown in Fig. 4 (*a*), which indicates the tumbler and sear when the lock is at full cock.

The trigger is merely a lever which is situated under the sear tail, and when the trigger is pressed it lifts up the sear tail, thus dropping the sear nose downwards out of the bent. When this happens the tumbler is no longer held in position; the main spring asserts itself and rotates it, together with the tumbler axle, with force and rapidity until the limbs are in the positions shown in Fig. 4 (*b*), where the sear tail has been lifted by the trigger.

The hammer is fixed to the tumbler axle, and so rotates with it. This rotation is termed the "fall" and is usually through an angle of about 30 degrees. At the end of its fall the hammer strikes the firing-pin, or striker, which in its turn hits the cap, and so causes discharge.

When the lock is cocked the hammer is pulled back, thus rotating the tumbler axle and tumbler in the reverse direction against the main spring, and when the rotation is complete the sear nose slips into the bent owing to the pressure of the sear spring on the sear tail. In fact, the lock is then "wound up" again.

All locks work on this general principle, although in bolt-action weapons the tumbler is replaced by a unit which slides backwards and forwards instead of one which rotates. But the principle of the bent and sear remains the same.

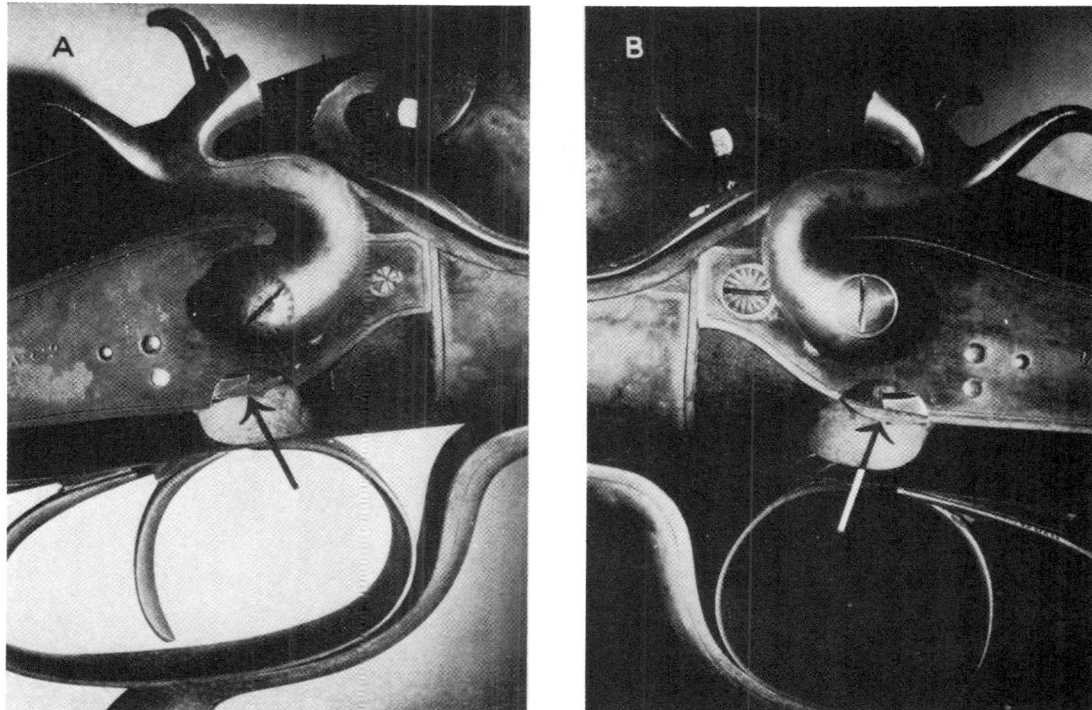

(A) The Right Lock of a hammer Sporting Gun with all the parts properly adjusted

A part of the lock plate and the wood of the stock have been cut away so as to render the sear nose and bent visible. The lock is at full cock, and the sear nose fits correctly in the bent (see arrow)

(B) The Left Lock of the same Gun, also at full cock

In this lock the sear tail is being forced up by the pressure of the wood of the stock with the result that the sear nose has been forced downwards and is half out of the bent (see arrow)

Plate X (*a*) shows the right lock of an ordinary hammer gun in the full-cock position. A portion of the plate which carries the lock mechanism (termed the "Lock Plate" or "Side Plate") has been cut away as has a bit of the wood of the stock into which the lock plate fits. This has been done to show the sear nose in position in the bent.

Now it will be realised that there are two forces tending to keep the sear nose in the bent: the pressure of the sear spring on the sear tail, and the pressure of the bent against the sear nose which is due to the main spring.

Of these two forces the former is comparatively feeble, and the latter depends on the way the sear nose fits into the bent. That is on: the depth of the bent; the shape of the bent; whether the sear nose fits right home in the bent or not; and the relative positions of the centre of the tumbler axle, the sear nose and bent in the full-cock position, and the sear peg.

All these points need attention when considering whether any particular weapon is abnormally liable to accidental discharge, but it can generally be assumed that in a gun *of proper construction* the easier it is to pull the trigger and fire the gun the more liable that gun is to accidental discharge.

For if the trigger pull is very light it must mean that the forces holding the sear nose in the bent are not very strong, in which case a sudden knock or blow will easily jar the sear out of position, and the gun will be discharged. And this is one of the reasons why a very light trigger pull is commonly, and quite rightly, considered dangerous.

Hammerless guns are fitted with a safety slide, and

there is a widely held belief amongst sportsmen that the operation of this slide actually cocks and uncocks the locks. In reality it does nothing of the sort, but merely bolts the triggers and so prevents them from being pressed against the sear tails. A safety device is unquestionably an extra element of safety, but when a gun is at "safe" it is still quite possible for it to be jarred off, although it is not so fully prone to accidental discharge as it is when the slide is not at "safe."

This liability to accidental discharge was recognised in the early days of hammerless guns and resulted in the invention of various "automatic intercepting safety" devices. I will not enter into any details of such mechanisms here, merely indicating that their object is to prevent the tumbler rotating unless the trigger has been actually pressed. In theory these intercepting safeties are perfect, but in practice it is to be feared that they are not. At the same time they undoubtedly do provide an extra element of safety, and on this account they are almost invariably fitted to all good-quality hammerless guns and rifles.

Hammer guns are seldom now fitted with the special safety stops which used to be seen on double hammer big game rifles and which certainly did lock the hammers when in the full-cock position with great efficiency. Consequently a hammer gun at full cock is less safe than a hammerless gun from the point of view of accidental discharge.

Similarly all rifles, pistols and revolvers are liable to accidental discharge when at full cock. Many of these weapons are fitted with safety devices which vary in degrees of efficiency, some being undoubtedly so effective as to render the weapon as safe as any *loaded* weapon can ever be.

But if the safety device is not in action pistols and revolvers are just as liable to accidental discharge when at full cock as any other type of weapon.

It can, in fact, be assumed that in actual practice no weapon can be regarded as immune from the possibility of accidental discharge if it can conveniently be fired by pressing the trigger in the normal way.

This should never be forgotten, and if any accident or fatality occurs owing to an alleged accidental discharge and suspicion falls on any individual the weapon in the case should be examined carefully and thoroughly in order to make sure that it is in good order, as any maladjustment of a lock mechanism can easily render a weapon peculiarly liable to accidental discharge.

And in this connection it may be of interest if I give some brief details of a recent case, especially as there can be no better instruction than a study of different methods of investigation.

A smallholder and his grown-up son were working together in the yard of their farm. The smallholder had a borrowed gun with him which he loaded, cocked and leant against the wall of one of the buildings so as to be ready in case any geese came over, as they frequently did at that time of year. Of course such carelessness is unpardonable, but it is unfortunately all too common.

The smallholder's story, which he told immediately to two neighbours to whom he ran for help and which he never once changed in any single detail, was that his son was shovelling some refuse within a few feet of where the gun was standing while he himself went to let their dog, a big heavy retriever, out of its shed. He did this and the dog rushed round the yard in excitement, and knocked the gun over. The gun was discharged and the son was shot.

After some delay the smallholder was arrested for murder, the case for the prosecution being that the gun was in perfect condition and could not possibly have been discharged accidentally, while the direction of the wound was such that the gun must have been in someone else's hand when fired.

No sort of evidence of motive was forthcoming, and the nearest approach to any serious quarrel had occurred four years before.

Now one of the commonest tasks which gunmakers are asked to perform by their customers is to overhaul their guns. Many shooters send their guns to their gunmakers at the end of every season for a periodic overhaul, and the gunmaker naturally removes the locks and generally takes the action to pieces in order to see whether every part is in proper condition and free from rust, and whether all screws are tight, and so on. This is normal business routine, as it is utterly impossible to form any opinion as to the condition of the internal mechanism without examining it.

It is also by no means a rare occurrence for some shooter to ask his gunmaker whether his gun is in its normal safe condition or whether it has in any way become abnormally prone to accidental discharge. In such circumstance the gunmaker *must* take off the locks of the gun, as it would be impossible for him to form any useful opinion without examining their mechanism.

At the first day's hearing before the magistrates,

however, cross-examination elicited the fact that the Prosecution had formed their opinion that the gun was " in perfect condition " and could not be discharged accidentally, without ever having taken the locks off to examine them, and had apparently arrived at their conclusion by testing the weights of the trigger pulls. That of the right trigger was exceedingly heavy, so heavy in fact that this lock probably was immune from accidental discharge.

But it was the left barrel which had been fired, and the trigger pull of the left lock was much less, being 7 lb., which is certainly heavier than that of a good-quality gun, but not particularly heavy for the left trigger of a cheap gun, since the pull of the left trigger is usually heavier than that of the right so as to reduce the possibility of the left lock being fired accidentally at the same time as the right.

Further cross-examination brought to light the fact that these trigger pulls had not been measured with a dead weight, which, as will be seen later, is the only possible method of obtaining an accurate result, but with a spring balance.

It was at this point in the proceedings that I was asked by the defending solicitor to examine the gun with a view to giving an opinion on its safety.

From an inspection of the outside of the gun I noticed two conditions of the left lock which suggested something being radically wrong.

In any lock which is in proper condition there should be a slight, but perfectly distinct, feeling of play on the trigger when the lock is at full cock. This is purposely left by the gunmaker in order to allow for the possibility of any change in the wood of the stock due to swelling or shrinkage caused by damp or dryness, and is thus a very important safety factor. In the left lock of the gun in question, which was a very inferior-quality gun, having all the signs of the cheapest type of continental manufacture, the trigger was absolutely rigid when the lock was cocked. This must always be a dangerous state for any trigger to be in, as there is no " give " in the case of a fall or knock, and so the force of such a fall is transmitted direct to the sear and the risk of accidental discharge is much increased.

The second point was that when the left lock was in the " down," or what is commonly, but not accurately, called the " half-cock " position, the hammer could be pushed right forward against the striker and the gun could be fired.

In my brief description of the mechanism of a lock I purposely confined myself to the principles which govern the actual firing of a lock. Hammer guns, however, have an additional cog, or bent, in their tumblers into which the sear nose fits when the hammer is not at full cock. In early guns this position was known as the " half-cock " position, and constituted the safety device, as when a lock was at true half cock the gun was immune from accidental discharge (owing to the deepness and shape of the bent) while the hammer could not possibly be pushed forwards against the striker.

In modern hammer guns the hammer automatically rebounds back into this position after firing, and so the safety device is automatic. The safety bent in the tumbler is called the " rebound bent."

In cheap guns there is often slight play of the hammer in the rebound position because the sear nose does not fit quite snugly into the bent, and when this occurs the hammer can be pushed forwards towards the striker to a limited degree which varies with the amount of play between sear nose and bent. But the forward movement of the hammer is always limited, and not complete.

In the left lock of the gun in question there was no sort of limit to this forward movement of the hammer, although there was such a limit in the right lock. This fact also indicated that there was something radically wrong with the left lock.

Accordingly I took the locks off, which was only the work of seconds rather than minutes since it was merely necessary to remove two screws in order to lift the locks off the stock.

The cause of the trouble in the left lock was then immediately obvious. The limbs had not been properly adjusted and the tail of the sear protruded below the edge of the plate which carries the whole lock mechanism. Now the wood of the stock is recessed to accommodate this mechanism and the lock plate fits exactly into the mouth of this recess so as to render the gun weatherproof.

It will be perfectly obvious that if any part of the mechanism protrudes beyond the edge of the lock plate, either that part must be forced out of position when the lock is in position on the gun, or else the lock plate will not completely cover the mouth of the recess.

In this particular gun the sear tail of the left lock had to be lifted up by about $\frac{1}{20}$ of an inch in order to permit the lock being fitted into its recess in the stock, and the result was that the sear tail was pressing hard on the wood of the stock all the time. The effects of this pressure on the wood were threefold:

(1) The sear nose was lowered and so could not engage fully in the bent when the lock was in the full-cock position.

(2) Similarly the sear nose could not engage at all with the rebound, or safety, bent, and this explained why the hammer could be pushed right forwards without any sort of check when the lock was at "half cock."

(3) The fact that the sear tail was pressing hard on the wood of the stock meant that any blow on the stock would be transmitted direct to the sear tail.

The combination of the first and third of these effects obviously rendered the gun peculiarly dangerous and liable to accidental discharge.

The effects of this lifting up of the sear tail by the pressure of the wood can be seen in Plates X and XI. Plate X (*b*) shows the left lock of the same gun as that shown in Plate X (*a*), but in this case the sear tail has purposely been bent until it protruded beyond the edge of the lock plate. The effect of the tail being lifted by the wood when the lock is at full cock is clearly seen, as part of the lock plate and wood have again been cut away to show the sear nose and bent. A comparison of the two photographs will show that while the sear nose is right home in the bent in Plate X (*a*), it is only half into the bent in Plate X (*b*). And in this position it will clearly be much more easily knocked right out of the bent by a sudden blow.

Plate XI shows the effect when the lock is at "half cock." Here again the right lock (Plate XI (*a*)) is as it should be, and the sear nose is engaging with the rebound bent, while in the left lock (Plate X (*b*)) the nose has been forced down so that it cannot engage with

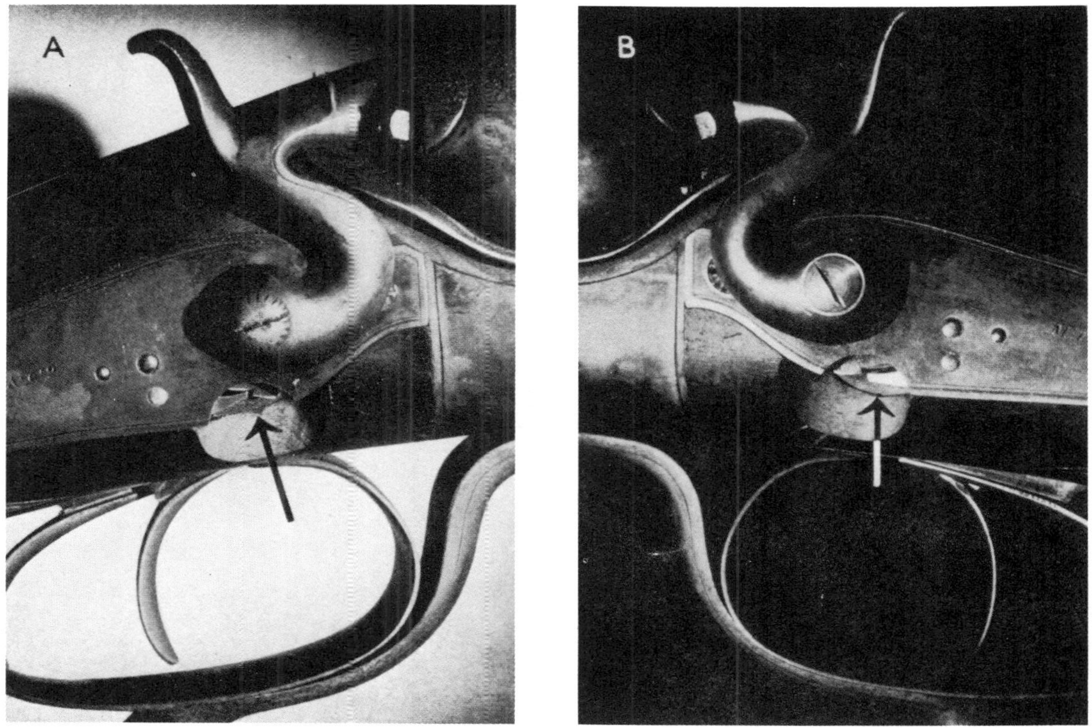

(A) THE RIGHT LOCK OF THE GUN IN PLATE X IN THE REBOUND POSITION
All the parts of the lock are properly adjusted and the sear nose is right up in the bent (see arrow)

(B) THE LEFT LOCK OF THE SAME GUN IN THE REBOUND POSITION
Owing to the sear tail being forced up by the pressure of the wood of the stock the sear nose has been forced down out of the bent (see arrow)

the rebound bent, with the result that the hammer can be pushed forwards without any check.

And there is the further fact that since the sear tail was resting on the wood it would be lifted up still more when the wood swelled in damp weather. Consequently tests for trigger pull after the gun had been kept for several weeks in a warm police station could hardly be expected to give a correct indication of what the trigger pull might have been at the time of the alleged murder, when the weather had admittedly been very wet.

In view of all these facts I formed the opinion that this particular gun was quite one of the most dangerous guns I had ever seen and that the left lock was peculiarly liable to accidental discharge.

This opinion was endorsed in every detail by three other expert witnesses, one of whom was Mr. R. K. Wilson, whose work on self-loading pistols has placed him in the front rank of authorities on the mechanisms of firearms, while the other two were practical gunmakers and directors of firms whose names are recognised throughout the world as the makers of guns and rifles of highest quality.

And it was one of these two last expert witnesses, who himself had had forty years practical experience at the bench and in the workshop, who stated in evidence at the Police Court: "If [the Home Office Expert] had only taken the trouble to undo two screws he could never have said this gun was safe. No gunmaker could."

Plate XXXVI shows the left lock of the same gun shown in Plates X and XI, in which the sear tail was bent down to protrude beyond the edge of the lockplate in a manner similar to that of the sear tail in the gun in the case. The effect of this is really obvious and can scarcely be regarded as a matter of opinion. The actual locks of the gun in the case are shown in Plate XII.

The evidence put forward by the prosecution in connection with the direction of the wound seemed, at the Police Court, to be based on the assumption that the dead man had been standing upright; but the Home Office Pathologist gave every appearance of confidence, as the following extract from a published report of his cross-examination by the defending solicitor will show:

> QUESTION.—" Would there be a great deal of speculation in the reconstruction of the tragedy?"
> ANSWER.—" Not necessarily."
> QUESTION.—" Do you say that there is no speculation at all?"
> ANSWER.—" If it is a matter of speculation it is not worth doing."

But when the case came up for trial at the Old Bailey this expert had seemingly changed his opinion, as the following extract from a published report will show:

> THE JUDGE.—" This case is all assumption, is it not? Is there any material upon which we can really draw a sound theory as to what happened?"
> WITNESS.—" The only material is that which shows the distance at which the weapon was discharged from the body, and the material which shows the direction in which the weapon was pointing in relation to the body."
> THE JUDGE.—" Your second point depends on the position of the body when the shot was fired?"
> WITNESS.—" Yes. The absence of scorching indicates that the gun was fired at some little distance from the body, I think probably not less than three feet."
> THE JUDGE.—" It is a ghastly speculation, the whole thing, isn't it?"
> WITNESS.—" There is nothing exact. It entirely turns on the position of the body at the time he received the wound."
> THE JUDGE.—" It is pure speculation as to how this man died?"
> WITNESS.—" Put that way, my Lord, it is."

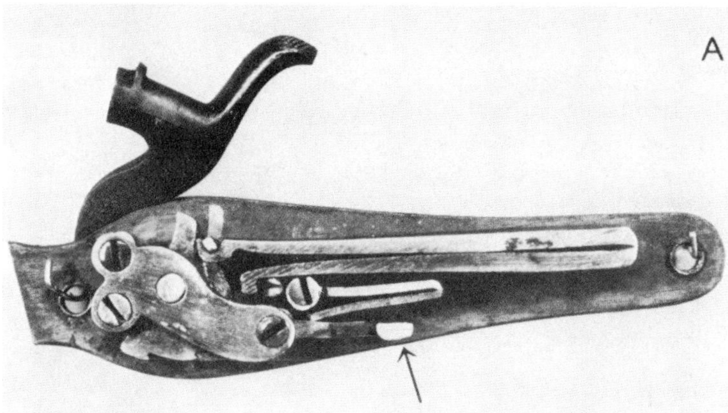

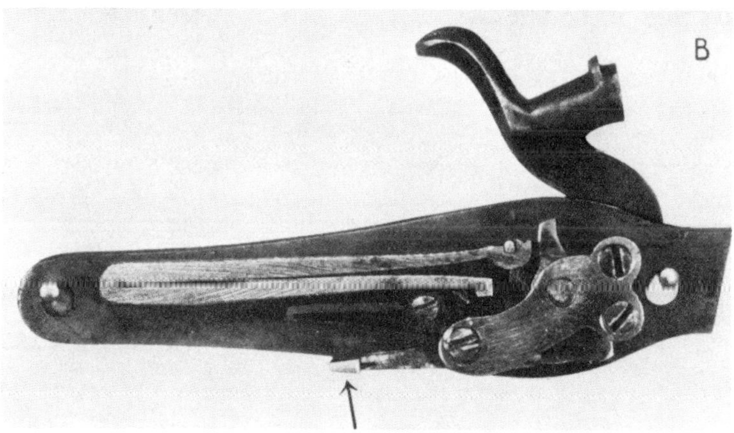

(A) The Right Lock of the Gun which was in the possession of a man accused of murdering his son

In this lock the various parts had been properly adjusted and the sear tail (see arrow) does not protrude below the edge of the lock plate

(B) The Left Lock of the same Gun

The sear tail hangs below the bottom of the lock plate (see arrow) and rested on the wood of the stock, thus rendering the gun highly dangerous and peculiarly liable to accidental discharge

After this no one in court was surprised when the case was stopped and the prisoner discharged.

I have purposely given the details of this case at some length in order to try to show how dangerous it may be to form an opinion of the condition of the lock mechanism of a gun without examining that mechanism, a fact which should really be obvious.[1]

So far I have only considered the possibility of an accidental discharge in the case of a weapon at full cock with the safety catch, if there is one, either on or off. But in all weapons fitted with external hammers, including revolvers and self-loading pistols, there may arise a question of an accidental discharge when the hammer is "down."

In a well-made hammer shotgun or rifle such a thing is, I think, seldom possible unless the blow is sufficiently violent to force the hammer forward against the spring which is holding it in the rebound position. And even then the strength of this spring would deaden any normal blow. But in a badly constructed gun it may easily be possible for the gun to be fired when the hammers are down, as has been seen. Every case must be examined with the greatest care and judged on its individual merits.

In revolvers and pistols, however, the matter is less simple.

In old-fashioned revolvers the danger of accidental discharge with the hammer down was so universally realised that no one with experience or knowledge

[1] I was so impressed by the mistakes made by the Prosecution in this case that I bought the gun in question and sealed it in a case in front of witnesses directly it was handed over to me and before it had been in any custody other than that of the Prosecution and Home Office. This sealed gun-case is still in my bank.

ever had a round in the chamber which was immediately under the hammer. Five chambers were loaded, but this one was purposely left empty for safety, and so the old-fashioned " six shooter " was really never more than a five-shot weapon.

In modern revolvers, however, fitted with rebounding hammers in which the firing-pin is withdrawn off the cap automatically after a shot is fired, there is much less risk. The best makes are fitted with an automatic stop which prevents the hammer from being forced forwards without the trigger being pulled. And I think that these stops are so effective that such revolvers can be regarded as being almost immune, if not quite, from a liability of being fired by a blow on the hammer.

And the same applies to the best makes of self-loading pistols which have external hammers.

The simplest way to test one of these weapons in this respect is to clamp the weapon with its barrel pointing vertically downwards in some sort of tube or pipe. An empty cartridge case with an unfired cap should be placed in the chamber immediately under the hammer and the hammer left in the ordinary rebound position. Weights can then be dropped on to the hammer so as to see whether it can be forced forwards so as to fire the cap.

Of course if two persons are struggling for the possession of a loaded revolver it can easily be fired by accident, because most modern revolvers have what is called a " double action " as well as the ordinary " single action."

In a single-action revolver the hammer must be cocked by hand before the weapon can be fired by pressing the trigger. But in a double-action revolver the hammer can either be cocked by hand, or it can also be cocked by a long pull on the trigger. Thus in order to fire all the cartridges in the cylinder it is merely necessary to pull the trigger right through after every shot, when the hammer is cocked and released by one long pull.

Hammerless revolvers have no single action, as the hammers are hidden within the stock, and the best makes of hammerless revolvers are fitted with safety devices. But even with these there must always be grave risk of a revolver being discharged by accident if two people are sufficiently unwise to struggle for its possession.

TESTING TRIGGER PULLS

It may frequently happen that a test is necessary to determine the weight of the trigger pull of some weapon, and consequently a knowledge of the correct method of making such a test is essential.

The commonest method is to use a trigger tester which is made in the form of a spring balance. Such testers are light, portable and convenient, and are consequently used largely by amateurs as well as by gunmakers. They serve their purpose well enough if an approximate reading is all that is required, but for an accurate reading they are valueless.

The only accurate method of testing a trigger pull is by means of a dead weight, and special trigger testers with variable weights form an indispensable item in the equipment of every armourer's shop, as well as of the leading gunmakers.

The futility of the spring tester is so well known that anyone who produced one at Bisley or any Small Bore Rifle Competition would be greeted with derision. In all competitions no test other than a dead weight is ever accepted. Consequently any evidence based on a spring balance which is put forward in a capital charge can only reflect very badly on the conduct of the case.

When testing for trigger pull the weapon should be cocked and the "finger" of the tester hooked on to the trigger. The weapon should then be held at such an angle that when the weight hangs vertically it will exert a pressure on the trigger at the same angle as that adopted by the finger when firing in the ordinary way.

The weapon should then be raised slowly until the weight is lifted off the ground. If the lock is not fired the test can be repeated with additional weights until the lock is fired. This gives the correct weight of trigger pull provided the weights are increased by not more than 4 ounces at a time. Several repeats should be made to check the value.

The angle at which the barrel of the weapon is held when making the test is of vital importance, and this varies with the type of action employed.

The following values for the angle of the barrel with the vertical may be taken as a guide when testing different types of weapons:

Weapon.	Angle of Barrel with Vertical.
Ordinary double shotguns and rifles (right trigger)	15 degrees
Ordinary double shotguns and rifles (left trigger)	25 ,,
Bolt-action military rifles	25 ,,
Martini-action rifles and most single-barrelled shotguns	25 ,,
Winchester Model 52 rifle	Barrel vertical
Revolvers	45 degrees
Single-shot pistols	45 ,,

It is impossible to lay down any definite angle for self-loading pistols as this angle must obviously depend on the relative positions of barrel and stock; and in self-loading pistols there is great variation in the inclination of the grip to the barrel. It is, therefore, probable that the best plan to adopt is to make the test with the stock horizontal.

The actual weights of trigger pulls in normal use do not vary greatly in the case of good-quality weapons, but in the case of cheap weapons they are generally somewhat heavier. This is because it is not possible to give the same care and finish to the locks of cheap weapons, with the result that these locks lack the stability of better mechanisms, and consequently the pull is increased in weight so as to provide the same margin of safety normally found in a good-quality lock.

The following weights of trigger pulls may be taken as a guide to usual custom in high-grade weapons, and any trigger pull much lighter than that quoted for a particular type of weapon should be considered as being potentially dangerous from the point of view of liability to accidental discharge. At the same time it should be realised that a heavier pull does not necessarily confer additional safety, as this depends on the actual finishing and adjusting of the limbs of the lock, as will have been appreciated from the description of the gun in the case which has been described

Weapon.	Normal Trigger Pull.
Ordinary shotgun (right trigger)	$3\frac{1}{2}$ to 4 lb.
Ordinary shotgun (left trigger)	4 ,, $4\frac{1}{2}$,,
Service Rifles	6 ,, 7 ,,
Sporting rifles	5 ,, 6 ,,
Miniature rifles	3 ,, 4 ,,

Weapon.	Normal Trigger Pull.
Pistols	3 to 4 lb.
Revolvers (single action)	3 ,, 5 ,,
Revolvers (double action)	16 ,, 20 ,,

The weight of pull given for double-action revolvers may, at first sight, seem terrific. But it must be remembered that this is quite a different type of pull to an ordinary trigger pull, as the trigger is gradually pulled back through a long traverse in order to cock the hammer and rotate the cylinder. The actual force required to pull the trigger is, therefore, distributed throughout a prolonged movement, and consequently does not appear to be intense at any particular moment during that movement. This is an experiment which anyone can try for themselves by snapping an empty revolver.

In cheap weapons the pulls are frequently from one to two pounds heavier.

CHAPTER VI

THE IDENTIFICATION OF FIREARMS BY MEANS OF FIRED CARTRIDGE CASES

IN the great majority of cases of murder by shooting there is seldom much doubt as to the weapon which was used, but there are occasions when the definite identification of some particular firearm forges the final link in the chain of proof. In this country the most famous example of this type of evidence is the Gutteridge case, in which the facts were briefly as follows.

In a motor-car which had undoubtedly been used by the murderers of P.C. Gutteridge the police found a fired revolver cartridge case. After months of work they were convinced that Brown and Kennedy were the murderers and found two revolvers in Brown's possession. But the link connecting Brown with the car was still missing. If it could be established that the fired cartridge case already mentioned had actually been fired by one of Brown's revolvers, and could not have been fired by any other, this link would be complete.

This was the first occasion on which the task of identifying individual firearms by means of fired cartridge cases or bullets was attempted in this country, and the Home Office very wisely sought the help of the War Office. The result was that the technical staff of Woolwich and Enfield conclusively proved that the fired cartridge case found in the abandoned car had beyond doubt been fired by one of Brown's revolvers; and the evidence for the prosecution was thus completed. I will give more details of this case later and merely mention now that it has never been generally realised that this work of identification was carried out by the War Office, and not by any Home Office or Police experts; so it is but fair that honour should be given where it is due.

The only other case which has been heard in England in which this type of evidence was used was the case of the shot Cypriot doctor, which was heard in March, 1933. I will also allude to this case later, and have only referred to it now because during the hearing the task of identifying individual firearms was referred to as a "new science." And there is no doubt that it has not been studied very widely in any country in the British Empire, although it has been utilised regularly in America since about 1920.

The methods employed are based on the fact that it is humanly impossible to make any two articles absolutely identical when they are viewed under the microscope. Even the surfaces of pieces of metal which are made by consecutive cuts on the same machine are microscopically different because the edge of the cutter becomes blunted during every cut, and so there are minute variations in the marks left on the surfaces which are severed in succession; for it must be realised that such surfaces are never perfectly smooth.

Similarly surfaces which are finished by grinding, or hand filing, present an appearance not unlike that of

PLATE XIII

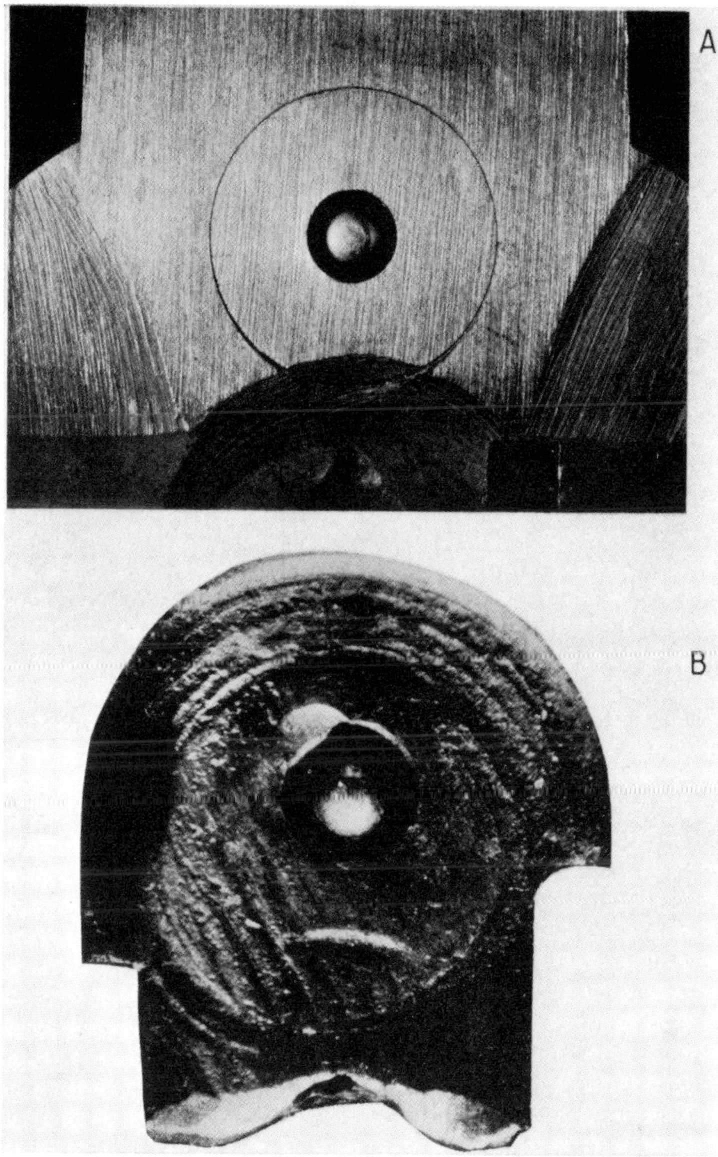

(A) The Breech Face of a high-grade, hand-finished Revolver

(B) The Breech Face of a machine-finished Self-loading Pistol

a ploughed field when viewed under a microscope, when the whole surface will be seen to be covered with minute furrows, or cuts, which have been left by the file or grinder. And just as a hard steel cutter becomes blunted during use, so do files and grinders become worn, with the result that they leave different marks behind them on the surfaces on which they are used.

Further, every stroke of a file will vary slightly in direction and force, which means that the direction and depth of the file marks on different surfaces will never be constant. And the same principle applies to surfaces which are finished by grinding.

The breech faces of all firearms are machine cut in the first place, and in the higher-grade weapons are finished off by hand filing or grinding. Similarly the strikers of all weapons are cut and shaped, receiving different degrees of fine finishing according to the grade of weapon in which they are to be used. Thus it will be realised that the breech face and the striker of every single firearm have microscopical individualities of their own. Sometimes, indeed, these individualities are so pronounced that they are visible even to the naked eye, while they can usually be seen with a good pocket lens.

Plate XIII shows the appearance of two typical breech faces when viewed under the low power of a microscope. The lower photograph is that of a machine-finished breech face of a self-loading pistol, on which the tool-marks are so obvious as to need no description. The upper photograph shows the breech face of a really high-grade revolver which has been finished by hand filing. Here again the markings are obvious.

But in addition to the actual tool-markings on a

breech face there may be other peculiarities. It may easily become indented by some knock or accidental touch with a tool even before the weapon ever leaves the factory, while such accidental markings commonly occur during use, especially if the weapon is not looked after very carefully. In fact, I am inclined to think that the two indentations indicated by lines in the photograph of the breech face of the self-loading pistol in Plate XIII were probably caused by blows with the end of a cleaning-rod.

Then in some makes of revolvers the shields (breech faces) are retained in position by a screw which holds them to the body, and the shield is consequently recessed to receive this screw. These recesses will vary in prominence and markings, just as the breech faces themselves vary.

When a cartridge is fired the pressure generated, which may be anything from 2 to over 20 tons per square inch, forces the case back against the breech face ; and since brass is softer than steel the individuality of the breech face is imprinted on the base of the cartridge. This imprint is received most clearly by the central cap, as the cap is always made of softer metal than the actual cartridge case.

In other words the firearm leaves its " finger-print " or " thumb-mark " on every cartridge which it fires. The distinctness with which this " thumb-mark " is imprinted naturally varies with the pressure ; but its image will always be present even though it is sometimes difficult to detect.

This is shown in Plate XIV, which gives photographs of the bases of an unfired revolver cartridge case and a similar case after it has been fired in a revolver. The

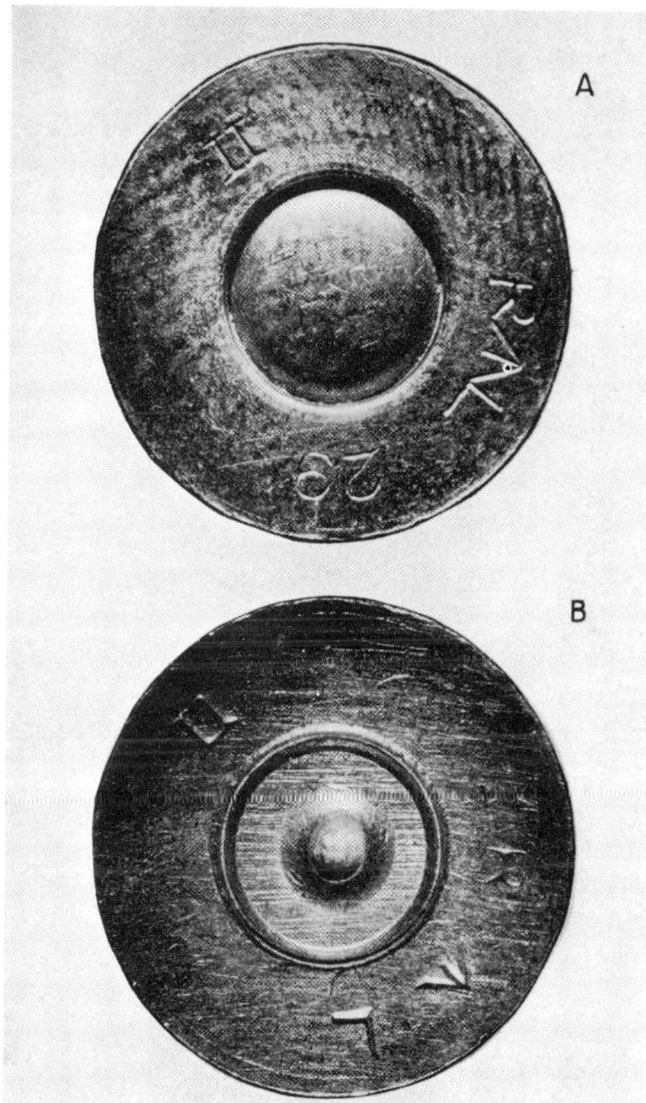

(A) An unfired ·455 Service Revolver Cartridge

It should be particularly noted that various markings are present on the case

(B) A fired ·455 Service Revolver Cartridge

The tool marks on the breech face have been impressed on the cap

striker indentation is obvious, while the striations imprinted on the cap are plainly visible.

Plate I shows how the markings on a fired cartridge case correspond exactly with those on the breech face of the weapon from which it was fired.

If there is any pronounced peculiarity in any particular breech face, this peculiarity *must* leave its imprint on the base of every cartridge fired from that weapon. An example of this is given in Plate XV, which shows three different fired cases all of which were fired from the same ·410 shotgun. The striker hole of this gun was flat on one side and *the imprint of this peculiarity is left on every cartridge case fired from that gun. So it will be realised that any ·410 cartridge case which does not exhibit this peculiar imprint could not possibly have been fired from the gun which fired the cases shown in Plate XV.*

The principle underlying this statement is one of the chief principles on which the identification of firearms is based.

In fact, the whole principle of identification is based on the fact that since the breech face of every weapon must be individually distinct, the cartridge cases which it fires are imprinted with this individuality. The imprints on all cartridges fired from the same weapon are the same, and those on cartridges fired from different weapons must always be different.

An example of this similarity is given in Plate XVI, which shows two cartridge cases both fired by the same revolver. The caps show fine striations, close together. But the upper portions have a pock-marked appearance owing to the breech face of the revolver being pitted by corrosion just above the striker hole.

It will now be instructive to examine cartridges fired from machine-finished self-loading pistols. Before we do so, however, it should be realised that self-loading pistols, and all bolt-action weapons, can leave other marks on fired cases in addition to the actual imprints of the tool and other markings of the striker and breech face.

It has been explained in Chapter I that all such weapons are fitted with an extractor claw (see Plate V) which slips over the rim of the case when the cartridge is loaded and withdraws the empty case when the breech is opened. This extractor is sometimes such a close fit that it produces a slight flattening of the rim when it rides over on loading while it often makes a slight scratch on the edge of the rim.

And in the same way the metal block which causes ejection (see page 26 and Plate V) leaves a distinct mark near the edge of the base, which is known as the "Ejector Mark."

This ejector mark is present in the great majority of cartridge cases fired from self-loading and automatic weapons and can be of the greatest help in the task of establishing the identity of an individual arm, as will be explained later. The nature of the mark varies greatly and depends on the shape and position of the ejector fitted in any particular weapon.

We can now consider Plate XVII, which shows three pairs of fired cartridge cases, which were fired from three different self-loading pistols, but all of ·250 (6·5 mm.) calibre.

The two cases comprising the middle pair were fired from an Astra pistol. The flattening of the rim caused by the extractor is very distinct: the ejector mark,

PLATE XV

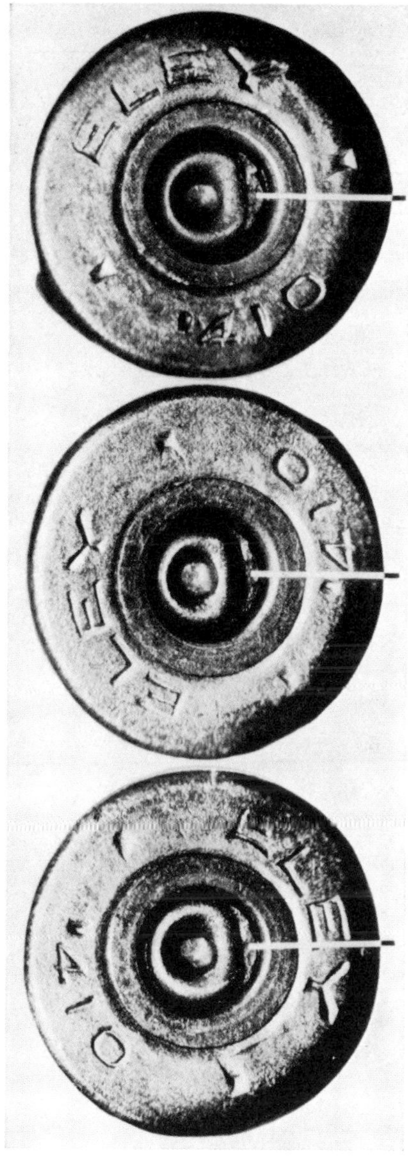

THREE DIFFERENT ·410 SHOTGUN CARTRIDGE CASES ALL OF WHICH WERE FIRED BY THE SAME GUN

The mark indicated by the line is present in all three cases

however, is faint, being merely an oblique but straight cut: but most prominent of all is the raised circular "tongue" on the cap, which is the imprint of a tool-mark. The combination of these three peculiarities, which together constitute the "thumb-mark" of this particular pistol, *must* be present in degrees of distinctness (which will vary with the pressure from round to round) on every single cartridge case fired by this pistol. If they are not present on a similar fired case, that case cannot have been fired by this particular pistol, and the upper pair of fired cases provide an illustration.

Here again these two cases were fired by an Astra pistol. The flattening caused by the extractor is similar, but the ejector marks are of quite a different type, while the tool-marks on the cap are equally distinct.

The bottom pair of fired cases were fired from the pistol of which the breech face is shown in Plates I and XIII. The constancy of the markings is plain, while the difference of these markings from those on either of the other pairs is obvious.

Plate XVIII gives yet one more example of three pairs of fired cases, but this time of rifle cartridges. The pairs in this plate were fired from three different ·303 Service rifles, and the identity of the markings of each pair is as clear as the difference of the markings of the different pairs.

Plate XIX shows six fired revolver cases, each of which was fired from a different ·455 Service revolver. These cartridges were purposely loaded in the revolvers with the "R" at 3 o'clock so as to obtain a comparison of the *direction* of the striations on each case, as well as of their nature.

But this plate also provides another lesson.

The centre of the striker is seldom absolutely concentric with the breech face and so in actual practice it will be found that in the great majority of fired cases the striker indentation is usually "ex-centric." This fact can, at times, have an important bearing on the identification of an individual arm, and so its existence should be noted carefully. The variation in the "ex-centricity" of the striker blows in the six revolvers used to fire the cases shown in Plate XIX is pronounced, even though they were all of the same manufacture (Royal Small Arms Factory, Enfield).

THE EFFECT OF PRESSURE

I have already mentioned in this chapter that the distinctness with which the thumb-mark of any breech face is imprinted on the base of a cartridge varies with the pressure. This is really obvious; but the obvious seems to be forgotten so frequently that it is impossible to emphasise too strongly the great importance of pressure in firearm identification.

It has already been explained in Chapter III that the higher the pressure the more the cap and base of the cartridge are flattened against the breech face; and it is this difference in degrees of flattening that can effect so materially the thumb-mark of any particular weapon.

This is probably best shown by an actual example, and Plate XX gives such an example. The two cartridge cases shown were both fired from the same ·32 revolver, but while the pressure developed in the upper cartridge was normal, that developed in the lower cartridge was comparatively feeble. This difference in

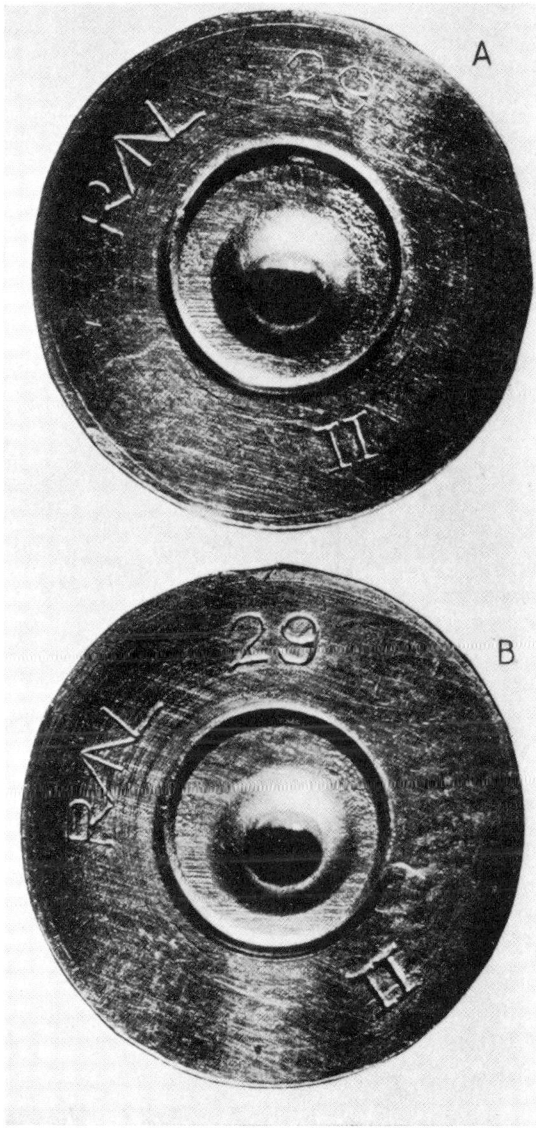

TWO DIFFERENT ·455 REVOLVER CARTRIDGES BOTH FIRED BY THE SAME SERVICE REVOLVER

The correspondence in the markings on the two caps is obvious

pressure can be detected at once by the difference in the degrees of flattening of the two caps. The flattened portion of the cap can always be recognised by the fact that it is the only part which is imprinted with any of the striations of the breech face. In the upper cartridge the greater part of the cap which is not actually occupied by the striker indentation is so imprinted, showing that the cap must have been well flattened against the breech face by a proper pressure. But in the lower cartridge only a narrow crescent of the cap is marked with the striations of the breech face, and the greater portion of the cap is still rounded.

An examination of the two cases will show that all the major markings on the upper case are also reproduced on the lower case, but not nearly so deeply. For the sake of clarity these different major markings have been indicated by numbered lines, the numbers corresponding in the two cartridges. The most noticeable difference is in the semicircular mark 4, which was caused by a recess in the breech face immediately below the striker hole. In the upper cartridge the cap has been flattened well into this recess; but in the lower the pressure was only sufficient to force the cap into the extreme end of the recess, while the main part of the recess, which was opposite the hollow surrounding the cap in the base of the cartridge, was never touched by the cap at all, and so left no imprint.

But there are two other lessons to be learnt from this plate. The first is that the very fine striations on the brass bases of the cartridge cases immediately *below* mark 3 do not run in the same directions in the two cartridges.

I attribute this to the fact that some of these finer

striations on the brass head were in existence before the cartridges were fired. I have frequently seen such markings on unfired cartridges, although it is more common to find markings in the form of concentric rings round the recess into which the cap fits. It must be remembered that when a cartridge case is impressed with the name of the manufacturer the die used for the stamping may have tool-marks on it, and that these tool-marks can be impressed on the brass of the base as well as the name of the manufacturer. But the die never touches the cap, and on this account it is seldom altogether safe to include small striations on the brass head of a cartridge as being part of the thumb-mark of a weapon unless it is found that they are reproduced in a number of cartridges fired from that weapon. Major markings, such as 1, 3 and 5, are quite reliable.

Another marking on the lower cartridge which might easily puzzle the inexperienced is 6. But if this mark is examined carefully it will be seen to comprise a cut on the edge of the brass and the cap which exists on the cap *in the hollow* between cap and cartridge base. Now it is quite impossible for any part of the breech face to touch this hollow, and so the mark 6 cannot possibly have been caused by the revolver, but must have been made either before or after the cartridge was fired. In this particular instance I happen to know it was before, because I fired, collected and examined these cartridge cases myself.

The second lesson is the importance of the circular hollow separating the cap from the brass base. Different manufacturers frequently fit caps of different sizes in cartridges of the same calibre. It may sometimes

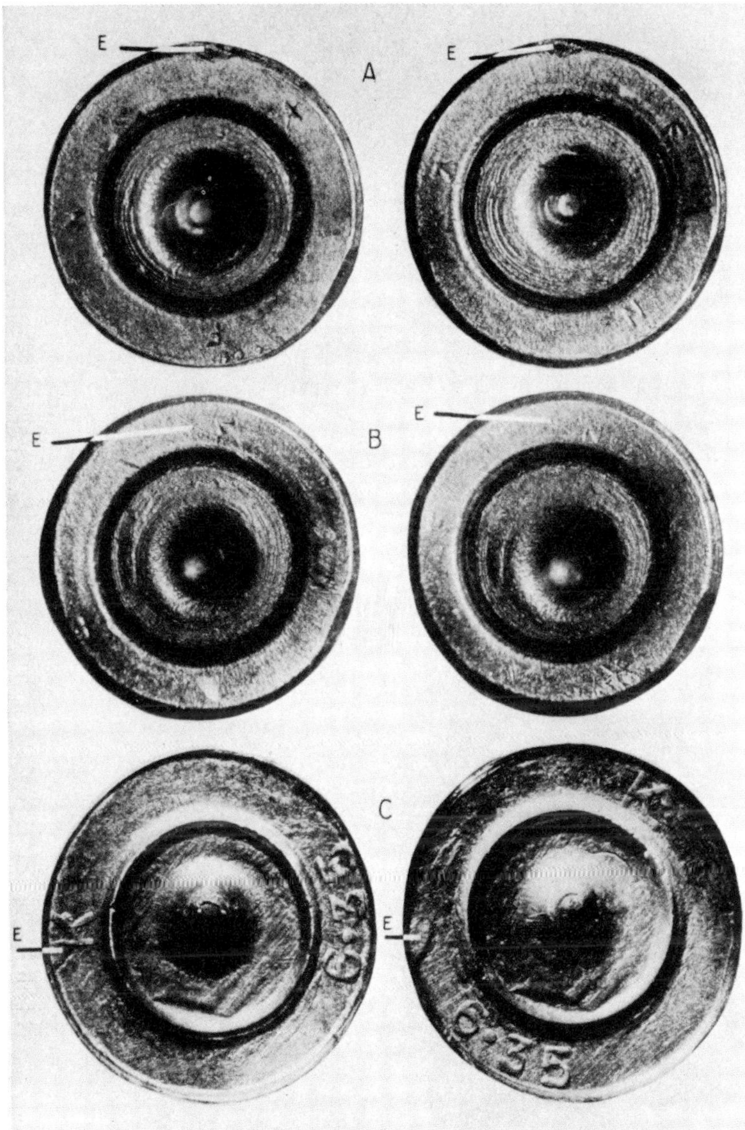

(A) Two different 6·35 mm. Cartridges, both fired by the same Astra Self-loading Pistol

In each cartridge the ejector mark is indicated by E

(B) Two different 6·35 mm. Cartridges, both fired by another Astra Self-loading Pistol

The ejector mark is again indicated by E

(C) Two different 6·35 mm. Cartridges, both fired by the same Mauser Self-loading Pistol

The ejector mark is indicated by E. These two cases were fired by the pistol of which the breech face is shown in Plate I, but they are different cases to the fired case shown in Plate I. The identity of the markings on all three cases fired by this pistol will be clear

happen that a very pronounced tool, or other, mark on a breech face happens to come immediately opposite this hollow in one make of cartridge, but just inside the hollow in another make in which a larger cap is employed. The effect would be that there would be no sign of this mark on a fired cartridge of the first brand, but that there would be a clear impression of it on the edge of the cap of a fired cartridge of the second brand. And the absence of the mark in the first cartridge might easily lead an inexperienced investigator to think that the two cartridges had not been fired from the same weapon.

In the cartridges shown in Plate XX the recess in the breech face comes very nearly opposite the hollow round the cap. But a proper pressure was sufficient to force the cap into this recess, and so a clear imprint was formed, as in the upper cartridge. But in the lower cartridge only the extreme end of the recess has just been reached by the cap, and its imprint was very nearly lost.

A final point which is worth noticing is the difference in distinctness of the two striker indentations. That in the upper cartridge is clean and sharp, while the imprint of the very nose of the striker has been obtained. But in the low-pressure cartridge in the lower photograph the whole indentation is ill-defined and " blurred," while there is no imprint of the nose of the striker.

THE STRIKER INDENTATION

The striker indentation can provide invaluable help in determining the thumb-mark of a weapon, although sometimes it is of little assistance.

It has already been explained that strikers, or firing-pins, are individually different and that their noses are covered with markings. These markings often take the form of a number of small concentric rings, and for this reason when the imprints of such rings are found in the striker indentation of a fired cartridge great caution is necessary in assessing their value as part of the thumb-mark of the weapon which fired that cartridge, because they may equally well form part of the thumb-marks of a host of other weapons.

Another by no means rare feature of a striker is the presence of a small " pimple " on the extreme end. Such a pimple leaves a minute indentation in the middle of the striker indentation, and an example is given in Plate XXI. The upper pair of cartridges were both fired from the same rifle, and both show the mark of the " pimple " on the end of the striker. In this case the presence of a similar mark on a similar fired cartridge would not necessarily prove that that cartridge had also been fired by the same rifle, unless other corroborating marks could be found. But if this " pimple " mark was absent from a similar fired cartridge, it would be proof positive that that cartridge could *not* have been fired from the particular rifle under consideration.

Similarly in the lower pair of fired cartridges shown in Plate XXI it will be noticed that both striker indentations are elliptical, owing to the rifle used having an elliptical striker. This in itself does not prove that all similar fired cartridges which might happen to have elliptical striker indentations *must* have been fired by the same rifle : but it *does* prove that any similar fired cartridge in which the striker indentation was not

PLATE XVIII

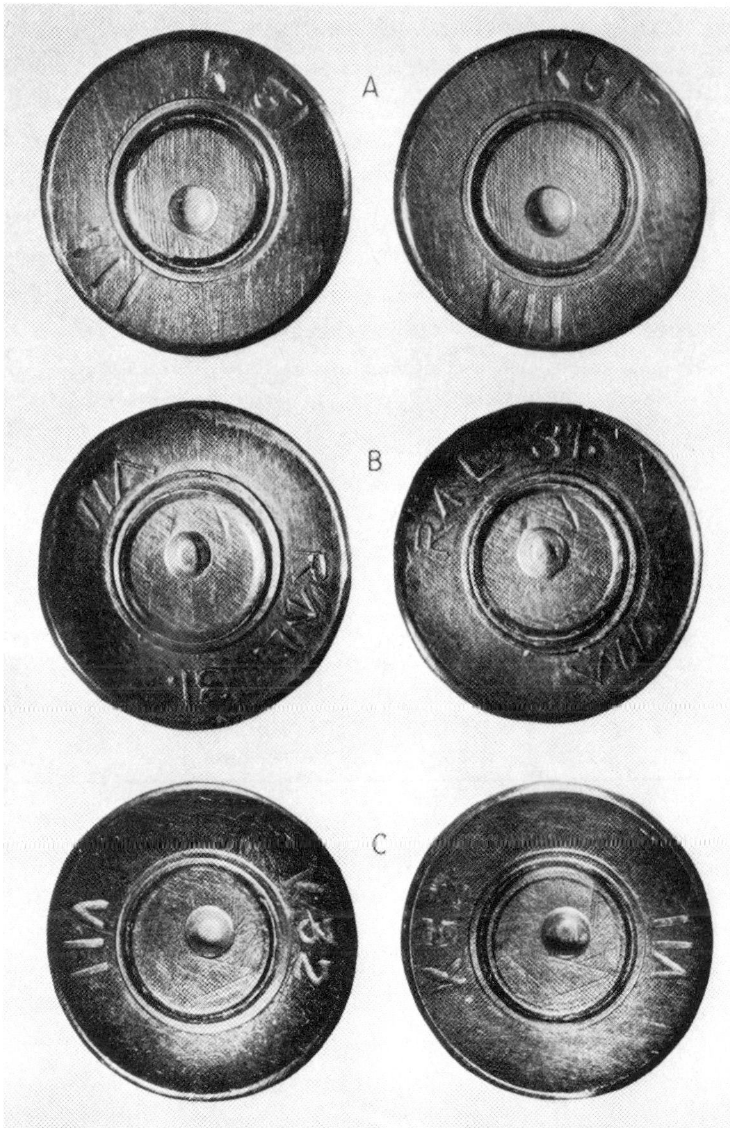

(A) Two DIFFERENT ·303 CARTRIDGES

Both fired by the same rifle

(B) Two DIFFERENT ·303 CARTRIDGES

Both fired by one rifle, but a different rifle from that used to fire the cartridges shown in A

(C) Two DIFFERENT ·303 CARTRIDGES

Both fired by the same rifle, but a different rifle to either used to fire the cartridges shown in A and B

elliptical could not possibly have been fired by that rifle.

So it will be seen that the general form of the striker indentation can provide useful positive evidence and conclusive negative evidence.

Another way in which the striker indentation can help is by its "ex-centricity." This point has already been touched on. In the case of revolvers the direction of the "ex-centricity" is always constant for any individual weapon, and the amount of "ex-centricity" fairly constant. That is, one particular revolver will always strike the cap at, say, 6 o'clock of the centre of the cartridge; but the divergence from the centre will not be absolutely constant because different makes of cartridges vary slightly in size, and so one cartridge may fit the chamber more tightly than another, and this will effect the degree of the divergence of the striker blow from the centre although not the direction of the divergence.

In self-loading pistols in which the firing-pin is a separate unit from the hammer and is free to slide backwards and forwards in a hole in the breech block the "ex-centricity" is not always necessarily constant because the actual direction of the forward thrust of the firing-pin will depend on the closeness of the fit of the pin in its guiding hole; and if the fit is loose, as it may quite easily be in cheap weapons, there is plenty of scope for lateral movement of the pin as well as longitudinal movement. While if the pin is itself not quite straight the direction of the blow which it delivers may vary to an appreciable extent from round to round.

And if the direction of the blow is not perfectly

straight the " ex-centricity " of the striker indentation may appear to be quite different in a high-pressure round from what it appears to be in a low-pressure round. This is because the indentation is much deeper in a high-pressure round, and so if the blow is slanting the divergence from the centre of the cartridge of the bottom of the indentation will vary with the depth of the indentation.

All the same, the " ex-centricity " of the striker indentation can be a very useful bit of subsidiary evidence even in cartridges fired from self-loading pistols provided its limitations are appreciated.

I have just stated that the depth of the striker indentation is dependent on the pressure. In fact the appearance of the striker indentation is one of the first points for which a trained ballistician would look when assessing the degree of pressure developed in a fired cartridge. Yet it has been asserted that the measurement of the depth of the striker indentation is one of the methods of determining whether two fired cartridges have both been fired by the same weapon. The best refutation of the entirely erroneous idea that a particular weapon always delivers a striker indentation of constant depth is Plate XXII, which shows two photographs of sections of cartridges both of which were fired by the same self-loading pistol. The striker indentation is clearly shown in profile in these photographs and the difference in depth is too obvious to need words.

As a matter of fact the depth is also dependent on the thickness of the rim in the case of revolver cartridges, and in the gauge and taper of the cases of self-loading pistol cartridges.

But the important point to realise is that the depth

of the striker indentation is useless as a means of identifying firearms.

Sometimes there occurs in a striker indentation what is known as a "striker scrape." An example of such a mark is shown in Plate XXII. This scrape is usually caused by slightly too long a striker. When the cartridge case is turned off the breech face on extraction there may not be sufficient clearance in the striker indentation for an abnormally long striker, and consequently the end of the striker scrapes the side of the indentation in the cap.

When a striker scrape is noted it can be of great help as it enables one to orientate the fired case correctly, that is to place it in the same position which it must have occupied in the chamber. It is always most useful to know, and frequently difficult to find out, how the cartridge fitted in the chamber, that is which way it was up. But a striker scrape enables one to settle this question at once.

It should be remembered that the orientation of a case by means of a striker scrape must depend on the type of weapon in which the cartridge was fired. For example, in an ordinary shotgun with a break-down action the striker scrape must always be at 6 o'clock. But in a self-loading pistol, or indeed any bolt-action weapon, in which the fired case is lifted away from the breech face in a different manner the striker scrape will correspond exactly with the ejector mark. That is if the ejector mark is at 8 o'clock the striker scrape will also be at 8 o'clock.

Sometimes a striker scrape is caused by an abnormally high pressure which so forces the cap back that it moulds itself round the striker which scrapes one side of the cap indentation when the fired cartridge is lifted off the breech face.

EJECTOR MARKS

I have already referred to the ejector marks which are usually present on cartridges fired from self-loading and automatic weapons. In some weapons the firing-pin is used as an ejector, and in others the next cartridge in the magazine or a part of the magazine, and in weapons of these types there will be no ejector mark.

When an ejector mark does exist it is, as a rule, obvious. It may be anything from a thin longitudinal cut to a big gash or indentation in the edge of the case. In some pistols the ejector is rounded and the mark is extremely difficult to detect unless the light catches it at exactly the right angle. So it is unsafe to assume that no ejector mark is present because one was not immediately visible; and in such circumstances a most careful examination of the base of the fired case should be made under a low power of the microscope while the case is slowly rotated through a complete circle so as to ensure the light striking it at every possible angle. It frequently happens that when a fired case is examined thus no ejector mark is visible until suddenly it stands out clearly and distinctly. But when the case is rotated a little more the mark vanishes once again.

The ejector mark, when present, forms a valuable bit of evidence, not only as part of the thumb-mark, but also as a means of orientating the case. For when two fired cartridge cases are being compared it is essential that they should both so be orientated that they must have been the same way up when in the

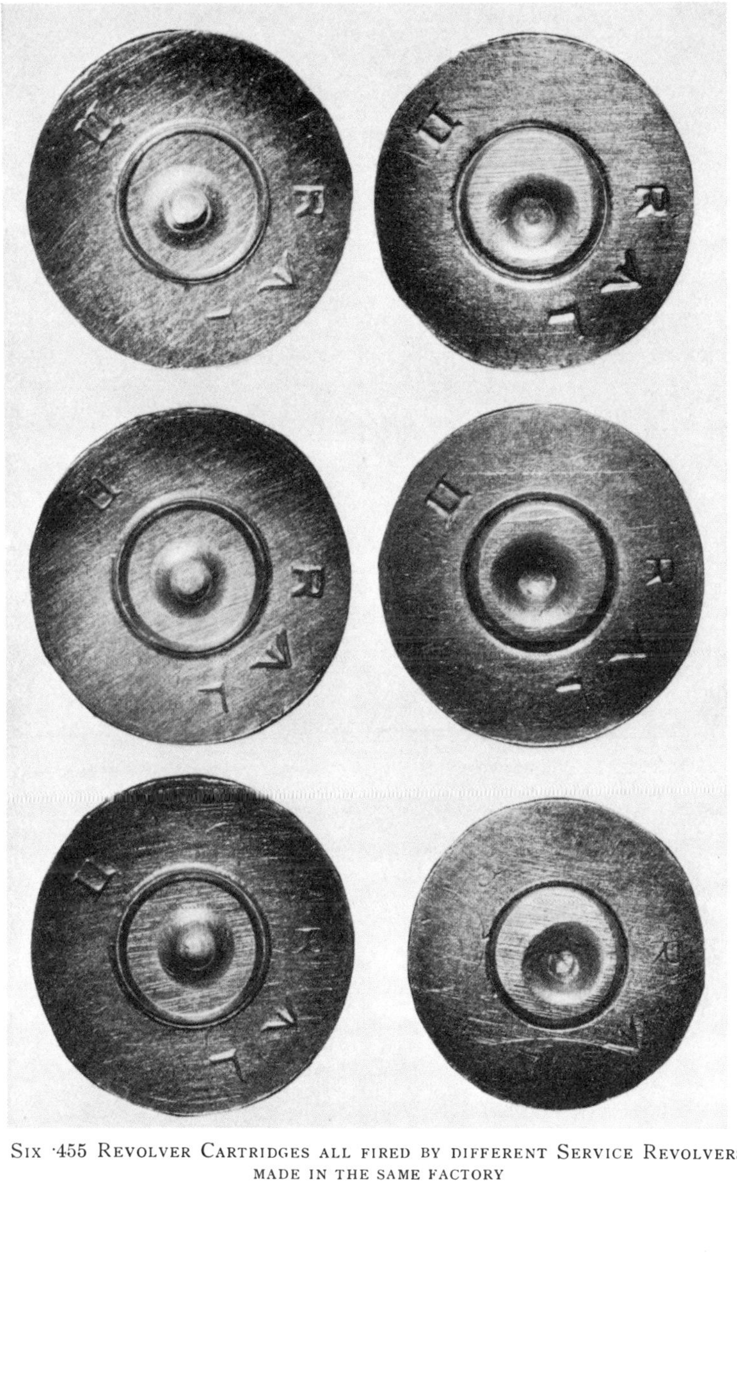

SIX ·455 REVOLVER CARTRIDGES ALL FIRED BY DIFFERENT SERVICE REVOLVERS MADE IN THE SAME FACTORY

chamber of the weapon. The ejector mark enables this orientation to be effected at once, and if two cartridges are to be compared and both have ejector marks they should both be placed with these marks in similar positions, that is both ejector marks at 3 o'clock, or 9 o'clock. The exact position does not matter provided the same position is utilised for both ejector marks. When this has been done the two cases can be examined with a view to finding other major marks common to both, and in the same relative positions in both cases.

Like the striker indentation the distinctness of the ejector mark is affected by pressure, even if indirectly. A high-pressure round causes the breech block in a self-loading pistol to be thrust back with greater violence, which means that the fired case (which is carried back with the breech block) is brought up against the ejector with greater violence, and so the resulting mark is more distinct.

But a weak round which is only just lively enough to push back the breech block will produce a very faint ejector mark.

MARKS ON THE SIDES OF FIRED CASES

So far we have only considered the marks on the base of a fired cartridge case. But there are sometimes marks on the side of a case which, if not so important as those on the base, can be of help in the task of identifying an arm. So it is essential that their usefulness and their limitations should be appreciated.

These marks can really be classified under two main headings: "Extractor Marks," and "Scrapes."

"Extractor Marks" are of two entirely distinct types. The first type is produced by the claw extractor to be found on all bolt-action weapons, and reference has already been made to them in this chapter. If the extractor fits very close it can so bruise the edge of the rim of the cartridge that it actually flattens the edge where it rides over the rim, as has already been seen. Such a flattening can be detected by looking at the base of the cartridge, but in the great majority of weapons the extractor does not ride sufficiently hard over the rim of the cartridge to cause any noticeable flattening, and it merely makes a little nick in the rim. These marks are extremely difficult to detect with certainty, as the rims of unfired cases are frequently so marked. But the extractor will also make a little mark on the forward edge of the rim when it pulls the fired cartridge case back with the breech block; and the combination of the mark caused by riding over the rim and that produced by pulling the fired case back may frequently be sufficiently pronounced to enable one to fix the mark with comparative certainty. But there can seldom be absolute certainty, for the extractor mark is produced by the act of loading and unloading, and is quite independent of the firing of the cartridge although it may be intensified in some weapons by the pressure developed on discharge. And how often are cartridges loaded in the chamber of some weapon only to be unloaded again without being fired. So it is perfectly possible for one cartridge to have a number of extractor marks on it, some even being caused by different weapons.

And since the identification of firearms should be carried out on exact lines, without any possible sources of error, my own personal opinion is that extractor marks should always be viewed with extreme caution unless they are so pronounced as to be clearly visible on the base of the cartridge.

The second type of extractor mark is quite different, and can be of great value, as it is only caused by actual firing. This type is only to be found in cartridges which have been fired with extractors, or ejectors, of the form commonly used in ordinary shotguns and revolvers, and can best be explained by an actual photograph. Such a photograph of a fired 12-bore shotgun cartridge is given in Plate XXI, and the extractor mark is indicated by lines.

The distinctness of this type of extractor mark depends on the pressure, as a high pressure expands the wall of the cartridge case until it receives an imprint of the junction of the extractor with the rest of the chamber. In fact one of the surest indications of a high pressure in a shotgun cartridge is the distinctness of the extractor mark.

The majority of revolvers are fitted with extractors which function on the same principle, and sometimes an extractor mark is visible on a fired revolver cartridge. But the wall of a revolver cartridge case is very thick near the head, and so the pressure is seldom sufficiently violent to expand this part of the case sufficiently to take an imprint of the extractor unless the extractor happens to be an abnormally bad fit. I have known a revolver which regularly produced an extractor mark on every cartridge which it fired. But the extractor fitted very badly, and such a weapon is, I think, quite exceptional.

In shotgun cartridges the brass heads are cupped,

and not solid drawn as are revolver cartridges, and so the brass at the head is quite thin and extractor marks are more often present than absent.

Similar extractor marks are also to be found on many ·22 rim-fire cases.

The real value of an extractor mark is that it enables one to orientate the cartridge case correctly. But it should not be forgotten that in double shotguns the marks on cartridges fired from the right and left barrels will occupy different relative positions.

" Scrapes " are comparatively common in cartridges which have been fired from self-loading or automatic weapons. When one cartridge is fired and ejected the next in the magazine is pushed up by the magazine spring and thrust forwards into the chamber by the breech block as it comes forward again after firing. If there is some projection, or even roughness, on the front part of the magazine or in the mechanism of the weapon the side of every cartridge may be scraped as it is pushed forwards into the chamber.

Similarly there may be a burr in the metal of the chamber itself which causes a scrape either as the cartridge is inserted into the chamber or else as it is withdrawn.

All scrapes of this nature which have been produced in one weapon will occupy a constant position on the side of the case relative to some fixed markings on the base, such as the ejector mark.

It is, therefore, useless to consider scrapes when working at an identification unless the cases are orientated by the base marks in the first place. If the scrapes on two or more cartridges then prove to occupy identical relative positions as well as being similar in appear-

PLATE XX

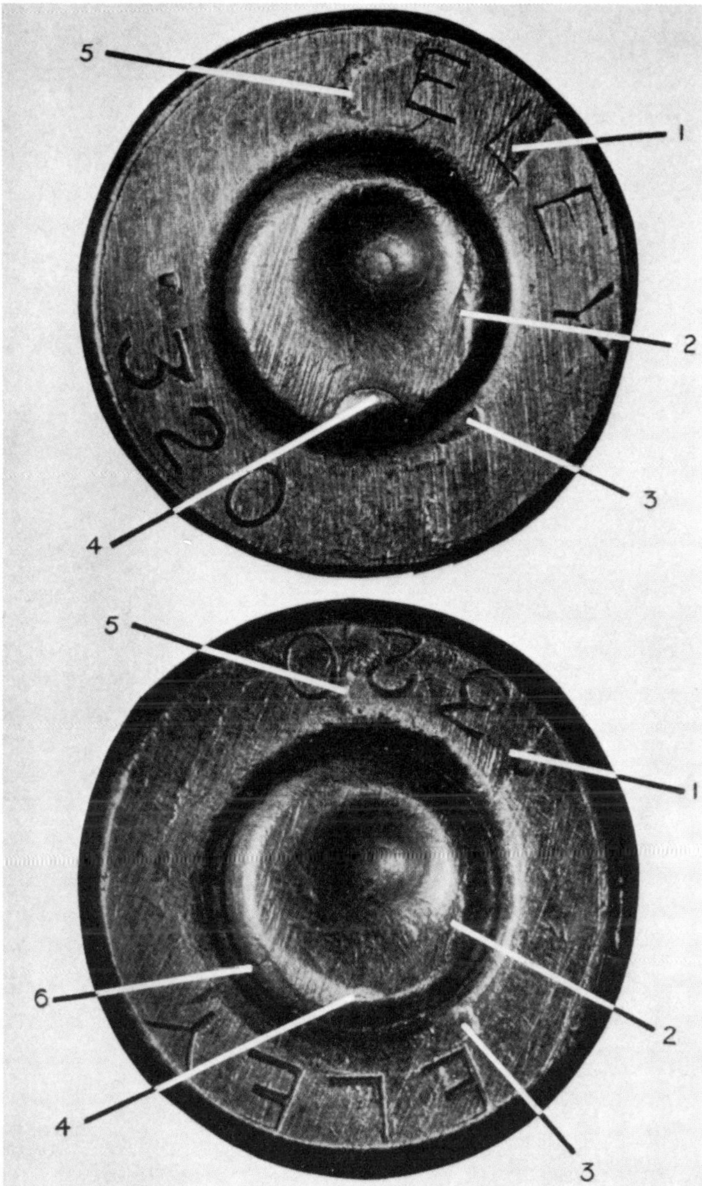

THE EFFECTS OF PRESSURE ON THE MARKINGS IMPRINTED ON THE BASES OF FIRED CARTRIDGE CASES

The pressure developed in the upper cartridge was normal and the various marks are clearly imprinted. In the lower cartridge the pressure was feeble, and the markings are much less strongly marked. But the two sets of markings will be seen to correspond

ance they can provide useful corroborative evidence of identification.

It must be realised, however, that all scrapes should be regarded with extreme caution because, like extractor marks, they can be produced during the process of loading and unloading without firing. Further, grains of grit or sand in the chamber or breech mechanism may quite easily produce scrapes on one or two rounds and then be removed, either by cleaning, or during the process of loading, firing, extraction and ejection.

Scrapes can also be produced in the actual manufacture of the cartridge case ; and although all cases with visible markings are usually rejected, some manufacturers are less particular than others.

Two very similar scrapes are shown in Plate IX. These are both fired ·250 self-loading pistol cartridges ; but not only were they fired from different pistols, but even from different makes of pistol. Yet I have actually seen photographs put forward as evidence for the identification of a particular pistol with the bullet extracted from a man who was shot, which showed scrapes almost identical to those shown in Plate IX. No attempt had been made to orientate the fired cases so as to show that the scrape occupied a constant position relative to some datum mark on the base of each cartridge, the evidence of identification being based on the fact that in each case the scrape was situated at approximately the same distance from the base of the cartridge. But the great majority of scrapes caused by burrs in the chamber are situated at approximately the same distance from the base, because the wall of the cartridge is not so thick nearer the top, and consequently the forward portion is expanded more readily by the pressure and so fits the chamber more tightly than the part nearer the base.

It may be suggested that the weak spot which I have emphasised in the evidence provided by extractor marks and scrapes, namely the fact that they can be produced without the cartridge having been fired, applies equally to ejector marks. But it does not. It is impossible to withdraw an unfired cartridge by hand in any self-loading or automatic arm with anything like the force which is exerted by actual firing; and it is the violence of the impact resulting from the discharge of the cartridge which produces the ejector mark on the base of the cartridge.

UNACCOUNTABLE MARKS

It sometimes happens that some very pronounced and seemingly distinctive mark is to be found on the base of a fired cartridge, and an inexperienced investigator will quite naturally think that he has here got an ideal thumb-mark. When he finds that this mark is not repeated on other cases which he has reason to suspect were fired from the same weapon he may easily be drawn to conclude that these other cartridges were actually fired by a different weapon.

Accordingly it should be realised that unfired cartridges are frequently marked heavily by careless loading and unloading, and particularly by careless insertion into a vertical magazine, such as is commonly used in self-loading pistols. These magazines all have the sides curved over at the top so as to hold the cartridges in position, and the front edges of these curved sides form lips under which cartridges must be pushed in

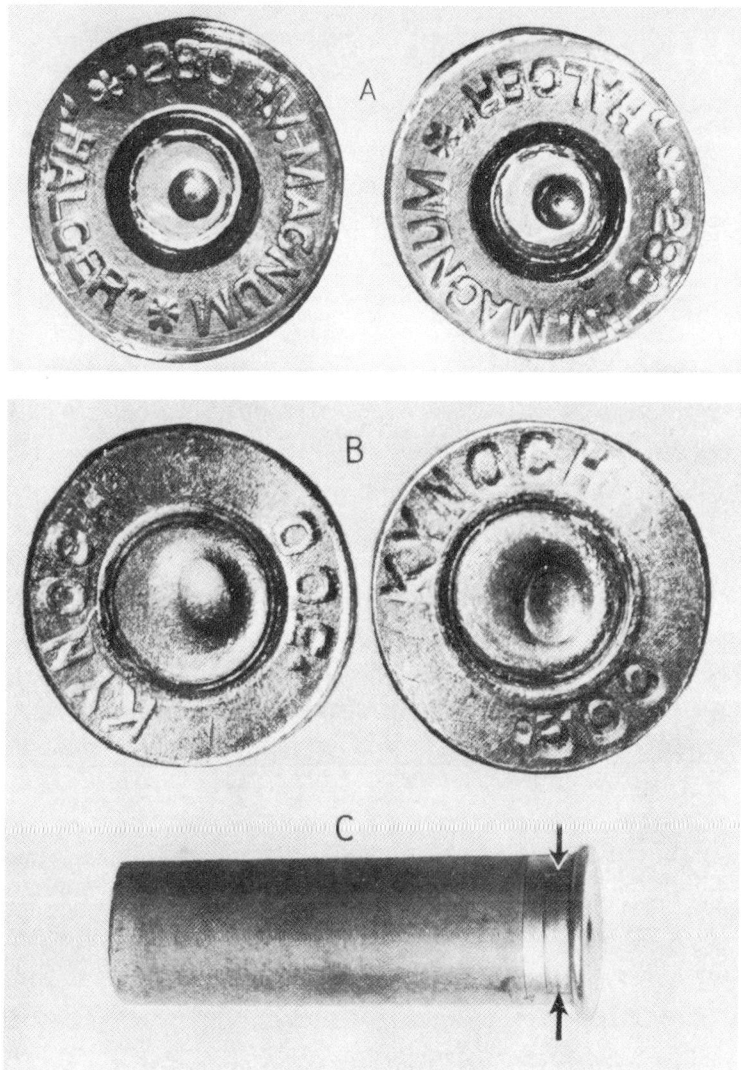

(A) Two different ·280 Cartridges both fired by the same Rifle

A small indentation will be seen in the middle of each striker indentation. These were caused by a small "pimple" on the nose of the firing pin

(B) Two different ·300 Cartridges both fired by the same Rifle

The striker indentation in each cartridge is oval

(C) A fired 12-bore Shotgun Cartridge

The extractor mark is indicated by the arrows

order to charge the magazine. If the person using the pistol is not accustomed to charging such magazines he will very likely score the base of almost every cartridge which he inserts into the magazine against these lips. When the cartridges are fired the bases will exhibit pronounced marks of this scoring, but the marks will differ in every cartridge, and this difference may sometimes be puzzling.

Plate XXIII (B) is a photograph of an unfired cartridge which has been scored in this manner. This particular cartridge was removed from the magazine of a self-loading pistol which belonged to a criminal, and it provides an excellent example of marks being present which are quite independent of firing and which are, therefore, useless as a means of identification.

This cartridge is also an example of an adaptation of a revolver cartridge so as to enable it to be fired in a self-loading pistol. The rim of the cartridge has been filed down, and this permits its usc in a self-loading pistol. This is a very common trick, and was used quite extensively in the Great War when revolver cartridges were more easily obtained than proper self-loading pistol cartridges. In England it is much more easy for a criminal to obtain possession of a pistol or revolver than of ammunition to fit that pistol or revolver; and consequently he may easily be compelled to make what improvisations he can.

Deep scratches on the base of a cartridge, and on the side too, can easily be made by unskilled charging of a magazine. Such a possibility should be realised, for it may quite easily happen that cartridges so scratched change ownership before being fired in a totally different pistol.

As a rule it is not a matter of great difficulty to detect the difference between marks which are imprinted on the base of a cartridge by the force of discharge and those which were there before discharge. It is probable that the majority of marks on the breech faces of weapons are *recesses*, in which case the imprint on the cartridge will be in *relief*. So any mark on the base of a cartridge which appears in relief must be caused either by an imprint of the breech face of the weapon which fired it, or else by the die used in stamping the cartridge with the mark of manufacture. Any marks which belong to this latter category will be very small and faint, and almost invisible to the naked eye no matter how the light is arranged to fall on the cartridge base. Marks imprinted by the force of discharge are invariably more pronounced, and in the great majority of cases there is no suspicion of any difficulty in distinguishing them.

But a mark caused before firing will always be a recess on the base of a cartridge, although it is perfectly possible for a recess to be imprinted by discharge. Doubtful marks should be examined closely to see whether there are any signs of firing-marks having been imprinted over parts of them, and when this occurs the matter is no longer in doubt.

To sum up it should be realised that marks produced before firing can at times be somewhat puzzling, but that in the great majority of cases they are very easy to distinguish, especially if the possibility of their presence is realised and their sources of origin understood.

PROCEDURE FOR IDENTIFICATION

I have now done my best to describe the various types of markings which may be found on a fired cartridge case and their sources of origin, and the next step is to consider how these facts can be utilised properly and correctly in the task of identifying some individual arm.

If a cartridge case is found on the scene of a crime, and a pistol is found later which may have been used by the murderer, the procedure to be adopted in order to determine whether the "crime" cartridge case "marries" the suspect pistol is as follows.

A series of rounds is fired from the suspect pistol the number constituting the series depending on whether any fired bullet has been recovered from the victim or not. If a bullet has been recovered, and it is found to be in sufficiently good condition to use as additional evidence for identification, the first series of rounds should not exceed five, for reasons which will be explained later. This series will presumably be fired by some investigator on behalf of the police. Later on, if a prosecution is decided upon, the defence will have the right of conducting an investigation of their own; in which case a second series of rounds will have to be fired. If bullet evidence is to be used the total number of shots fired through the barrel should not exceed ten, unless the evidence is to be of a very general character. It is for this reason that I have suggested that the first series of rounds should not exceed five if a bullet has been recovered.

But if no bullet has been recovered, or the bullet is too disfigured to be of any use in identification, the number of rounds is immaterial, and the more the better, up to a point, as the evidence may be strengthened.

At least two of these test rounds should be fired with " oiled cases." That is the sides of the cartridge should be well oiled before it is loaded in the chamber. The effect of this oil is to reduce the friction and so reduce the tendency of the case to stick to the inside of the chamber when it is expanded on discharge, and thus increase the force with which the base is driven against the breech face. This naturally means an increase in base pressure which tends to produce a more perfect imprint of the breech face on the base of the cartridge, and so helps the work of detecting the thumb-mark of the weapon.

After every shot the fired case should be carefully collected. This is not always too easy a matter, as some self-loading pistols eject their fired cases to an astonishing distance, while different makes throw the cases in different directions. It is, therefore, a wise plan to cover the hand and breech portion of the pistol with a rag or duster before firing, which will prevent the fired case from being ejected too far and lost.

The bases of all the fired cases should then be examined under a microscope in order to find out the thumb-mark of the suspect weapon. As has already been explained the thumb-mark of individual weapons will vary in form and size, and I have given a number of typical examples. I suppose that different individuals will naturally work on somewhat different lines, and so I will merely suggest the procedure which personally I find the most convenient, quick and certain.

For a good many years I have made it a habit to examine a few samples of any lot of fired cartridge

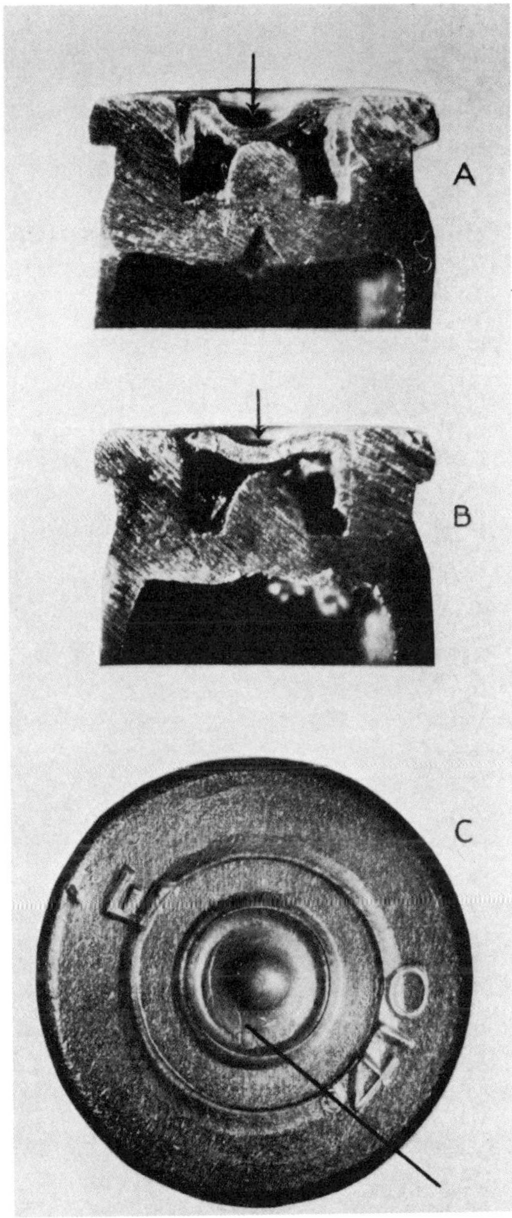

(A) A CROSS SECTION OF A CARTRIDGE FIRED BY A SELF-LOADING PISTOL

The pressure was normal and the depth of the striker indentation, which is indicated by an arrow, is also normal

(B) A CROSS SECTION OF ANOTHER CARTRIDGE FIRED BY THE SAME SELF-LOADING PISTOL

The pressure was low and the striker indentation (indicated by an arrow) is not nearly so deep as in the case shown above

(C) A FIRED ·410 CARTRIDGE SHOWING A "STRIKER SCRAPE"

cases which I happen to find. It may be a heap on the firing-point of a rifle range, or at some stand in a pheasant covert: I always examine a few with a ×10 aplanatic pocket lens which is my constant companion. I have thus examined many thousands of fired cases, and constant practice enables one to spot markings very quickly. All that this preliminary examination does is to show some constant and prominent mark by which it is possible to orientate all the test cases similarly.

Having done this I stick the test cartridges on an ordinary glass microscope slip with plasticine, with, of course, their bases upwards. They should be orientated similarly—this can be done quite easily with the pocket lens—and placed in a row as close to each other as possible.

I then examine them under a microscope, beginning with as low a power as possible, usually a 4-inch objective and a ×5 eye-piece. With this low power one can often get three cartridges in the field of view at a time, and the illumination is then adjusted so as to fall quite obliquely on the bases. I then slowly turn the cartridges through 180 degrees by means of the revolving stage of the microscope. This ensures the light striking them at every angle, and in the course of the rotation a point will come when some mark or marks suddenly shows up very distinctly. When this point is reached I look for this mark in all the cartridges which are in the field of view, and if it is present in all I bring the rest of the test cases into the field one by one by traversing with the mechanical stage, taking care always to keep a cartridge which has already been examined in the same field as a new cartridge, so as to ensure getting a guide in every case.

In revolver cartridges there will be no ejector mark, but orientation can be obtained just as well by the striations on the cap, that is by fixing the cases so that all the cap striations are running in the same direction.

If there are no very pronounced and isolated marks common to all the test cartridges it is a help to use a higher power of magnification, and a 3-inch objective can be substituted for the 4-inch. Even with one of these objectives it is possible to get the bases of two ·38 revolver cartridges in the same field simultaneously.

With this power a more detailed comparison can be made of the finer markings, particularly the cap striations, and there is seldom much difficulty in finding several characteristics in the markings which are common to all the test cartridges. Once these characteristics have been discovered the thumb-mark of the suspect weapon is known, and the next step is to examine the breech face of that weapon so as to check the thumb-mark. This can usually be done quite well with a good pocket lens, but a more leisurely examination can be made with a microscope if it is possible to fix the breech face on the instrument.

Having now settled the question of the thumb-mark beyond all doubt it but remains to examine the " crime " cartridge to see whether the thumb-mark is also reproduced on its base. This can be done most simply by placing the " crime " cartridge and a well-marked test cartridge side by side on a glass slip, orientating them similarly under a pocket lens by means of the ejector marks, cap striations, or any other convenient feature which may seem to be common to both. The two

(A) A FIRED 12-BORE SHOTGUN CARTRIDGE WITH A VERY PECULIARLY MARKED STRIKER INDENTATION CAUSED BY A DAMAGED STRIKER

(B) AN UNFIRED CARTRIDGE TAKEN FROM THE MAGAZINE OF A CRIMINAL'S SELF-LOADING PISTOL

This cartridge is really a rimmed revolver cartridge, but the rim has been roughly filed down so as to make the cartridge function in a self-loading pistol. The scoring across the cap should be noted, as this was caused by the lip of the magazine when the cartridge was carelessly pushed into the magazine

cartridges can then be examined in the same field of the microscope, when identity or difference should readily be established.

Sometimes, when the markings are very fine, it may be advantageous to use a higher power than the 3-inch, and then a 2-inch objective can be employed. With this the field is too limited to accommodate more than one small-sized cartridge, or the part of a larger one, at a time ; but by traversing with the mechanical stage the cartridges on the slip can be examined in turn and their similarities or differences easily detected.

A 2-inch objective may also be occasionally useful to determine the presence of some minor marking in a low-pressure round in which the imprint is not very clear. But even so the occasions on which it is required are few and far between, and a 3-inch objective will be found by far the most generally useful.

A higher power than a 2-inch should never be required for cartridge cases provided the illumination is correct. For illumination has far more to do with rendering detail visible than has magnification. But I propose to deal with this subject later on.

The sides of the cartridges can be examined by mounting them on a rotating spindle with plasticine at right angles to the axis of the microscope. The illumination is focussed obliquely on one edge of the side and this edge also focussed with the microscope. Any markings along that part of the edge will then be seen distinctly. The case is then rotated so that every part of the side can be examined in turn.

I have purposely explained the procedure of examination with some detail because the instrument usually adopted and recommended for this work is that known as a "Comparison Microscope." Such an instrument really consists of two microscopes, optically paired, and mounted parallel one to the other. One cartridge to be examined is put under one of these microscopes, and another under the other. The two microscopes are connected by a special eye-piece which enables the observer to view both cartridge cases, or portions of both cases, in the same field of view at the same time. He is thus enabled to compare the marks on the two cases just as one can compare two photographs which are placed side by side more easily than when they are seen one after the other.

The most fantastic claims have been put forward for this type of instrument which are enough to suggest that it has magical properties, and that it automatically, and wholly of its own accord, rings a bell or utters some similar warning, when the two cartridge cases under examination exhibit the same thumb-mark! Unhappily there is no foundation for this comforting belief. In the hands of a trained microscopist the comparison microscope can be of great value in determining the identity of fired bullets; but for cartridge cases I have come to the conclusion that a high-class single instrument is preferable. After all, the sole advantage of the comparison microscope is that it enables one to see two objects in the same field. But this can be done, as I have explained, quite easily with an ordinary microscope; and after exhaustive trial I have now given up my comparison microscope entirely for cartridge cases, keeping it only for the examination of fired bullets.

For cartridge cases I use a "Horizontal" microscope by Messrs. W. Watson. This form of instrument was originally designed for metallurgical work, and is peculiarly suited for examining cartridge cases and the breech faces of weapons, as it is fitted with both body and stage focussing, and there is an altogether exceptional clearance between stage and objective.

It is also an ideal instrument for photo-micrography, which must always play a very important part in the task of identifying firearms, as without good photographs evidence of identification must consist merely of statements of opinion, when its value is not great.

So far I have confined myself to what may be termed positive evidence of identification, that is the finding of the thumb-mark of the suspect pistol on the " crime " cartridge.

This fact in itself is not necessarily sufficient definitely to marry pistol and cartridge, because although it is true that every breech face has an individuality of its own, it is also a fact that all cuts made by the same tool in a machine will bear a strong " family likeness " to each other. And since brass is not a perfectly plastic substance, such as warm wax, it may happen that the cartridge case is only imprinted with the " family " thumb-mark common to every breech face of that particular batch, and not with the individual thumb-mark of the breech face of one single pistol.[1]

A rather low pressure, for instance, may easily result

[1] The existence of a " family likeness " between different cuts with the same tool was very properly used by the Crown in a recent prosecution for murder. Evidence was given that the surfaces of two severed telephone wires showed the same characteristics when viewed under the microscope as the severed surfaces of similar wire which was cut for test purposes, and from this it was deduced that the knife which was used for cutting the test wire had also been used for cutting the original wires.

in the cartridge being imprinted only with some "family" mark instead of the smaller individual marks of one pistol. Similarly the brass used in different makes of cartridges, as well as the metal used in the caps, may vary in hardness and so vary in their readiness to receive a clear imprint of the breech face. As a matter of fact cartridge brass is nearly always 70 per cent. copper and 30 per cent. zinc, although some makers adopt 71 and 29 per cent. while others adopt 69 and 31 per cent. But they all keep near the 70 and 30 per cent. The metal used in caps, however, does vary in its susceptibility to receive imprints, and for this reason the test cartridges should always, whenever possible, be of the same make as the "crime" cartridge.

But the possibility of the existence of a "family" thumb mark must be kept in mind, and before it is possible to declare that the "crime" cartridge marries the suspect pistol it is essential to prove that it does not marry equally well *any single one of a large number of similar pistols of the same make.*

In the case of hand-finished weapons where the evidence of identification is based on the identity of a number of small striations, the question of the "family" thumb-mark is not of great moment. But even in such a case the positive evidence of identification would be strengthened materially by the additional negative evidence that the "crime" cartridge did not marry any single one of, say, fifty similar weapons to the suspect weapon.

And when identification is based on comparatively vague generalities of major markings, as it sometimes is although it *never* should be, this negative evidence

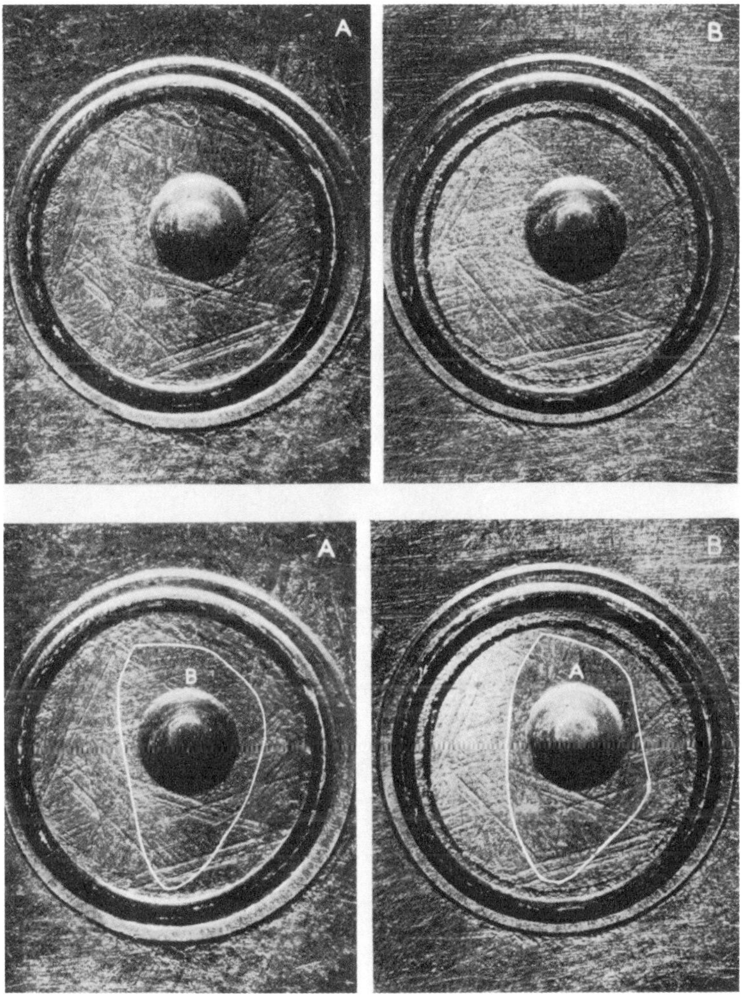

COMPOSITE PHOTOGRAPHS SHOWING HOW THE MARKINGS ON CARTRIDGES FIRED BY THE SAME WEAPON MATCH EXACTLY

A and B are the caps of two different ·303 cartridges, both fired by the same rifle. The lower photographs are both composite, that on the left being made up of A with the central part of B inserted; and that on the right being made up of B with the central part of A inserted. It will be seen that in both cases the markings on the caps correspond exactly

becomes essential and without it the positive evidence is completely valueless.

Let me give an example.

Some time ago I was engaged in a case which was heard overseas and in which the Crown expert based his identification of two fired cartridge cases found on the scene of a crime with a self-loading pistol found in the possession of the accused on the following: "identity" of ejector marks; "identity" of cap indentations; a single scratch on the upper part of the side of each cartridge.

What he meant by "identity" was never quite clear because he declared that there were no imprints of tool markings of the striker and breech face on either of the "crime" cartridges. And this in spite of the fact that the pistol was a cheap machine-finished weapon!

The photographs of the cartridge bases produced by the prosecution were ordinary photographs and not taken through a microscope and showed nothing beyond a similarity of ejector marks in the "crime" and test cases. But although different makes of pistols will give very different ejector marks, and although even different batches of one make of pistol will produce different ejector marks, there are almost bound to be a number of pistols of that make which will produce similar ejector marks. And yet not a single test had been made with any other pistol other than the one belonging to the accused.

The "identity" of the striker indentations was supported by the crudest attempt at photo-micrography which I have ever seen except one—and that was the photo-micrograph taken by the same expert in the same case of the "crime" bullet. This last photograph was merely a dark smudge, while the one of the striker indentations was a vague blur in which not one vestige of detail was visible.

The "scratch evidence" was that to which I have already referred on page 127.

Good photographs are an absolutely essential part of the evidence for the prosecution which is based on the identification of the accused's weapon with the "crime" cartridge. These photographs need not necessarily be taken at a high magnification—5 diameters is usually ample—but it is important that the photograph of the "crime" cartridge and those of all the test cartridges should be taken with the source of illumination at exactly the same angle in every instance, as a variation in the angle of lighting can easily render some important mark invisible.

If the identification is based on the smaller striations on the cap "composite" photographs taken at a higher magnification can provide most convincing evidence, especially to a non-technical jury.

An example of "composite" photographs is given in Plate XXIV. The top pair of photographs show the caps of the two ·303 cartridges which comprise the bottom pair in Plate XVIII, but it will be seen that the magnification is distinctly higher. The lower pair of photographs are each made up with the central part of one cap cut out and inserted over the other cap. But in spite of this "patchwork" arrangement the two lower photographs look but little different from the upper ones, while the various markings fit in perfectly in each case. These "composite" photographs are largely used in America where evidence of the identification of firearms is far more common than in this country.

CHAPTER VII

THE IDENTIFICATION OF FIREARMS BY MEANS OF FIRED BULLETS

SO far we have only considered the methods employed in the identification of individual firearms by means of fired cartridge cases. But it may easily happen that the only bit of material evidence available is the bullet which has been extracted from a victim, and in such circumstances the task of identification must be effected by marrying this bullet to a particular weapon.

The general principle employed is exactly the same as that used with fired cartridge cases. No two barrels are microscopically identical, as the surfaces of their bores all possess individual and characteristic markings. In fact every barrel has its thumb-mark in exactly the same way that every breech face has its thumb-mark.

When a bullet is fired from a rifled barrel it becomes engraved by the rifling, and this engraving will vary in its minute details with every individual barrel. So it happens that the engraving on a bullet fired from one barrel will be different to that on a similar bullet fired from another barrel. And conversely the engraving on bullets fired from the same barrel will be the same. So the task of identification depends on matching the engraving on two, or more, fired bullets in exactly the same way as it also depends on matching the imprints on the bases of two, or more, fired cartridge cases.

And since some readers may not be familiar with the appearance of the engraving on a fired bullet I have included photographs of unfired and fired lead and nickel-jacketed bullets in Plate XXV. In both cases the engraving consists of deep major furrows with fine striations in between these major furrows.

On each fired bullet the deep furrows have been engraved by the lands, and the striations in between these major deep furrows by the tool marks in the bottoms of the grooves.

It will be noted that only the lower portion of each fired bullet is engraved. This is only to be expected, as the parallel part of the bullet is the only part which takes the rifling, the tapered nose being of too small a diameter. But it will be seen that upper portion of the land furrow in the lead bullet has the appearance of being "doubled," this "doubling" decreasing as it approaches the base.

This is a very common feature in fired bullets and is caused by the bullet failing to enter the bore absolutely nose on. The result is that the forward part of the parallel portion of the bullet takes the rifling, and is engraved accordingly. But as the bullet travels farther into the bore, and the whole of the parallel portion enters the bore, any tendency to oblique movement is checked, and the bullet is forced into the rifling with its longitudinal axis coincident with the axis of the bore. When the bullet is so forced to assume true nose on movement it skids from its initial slanting direction and takes the rifling correctly in a manner

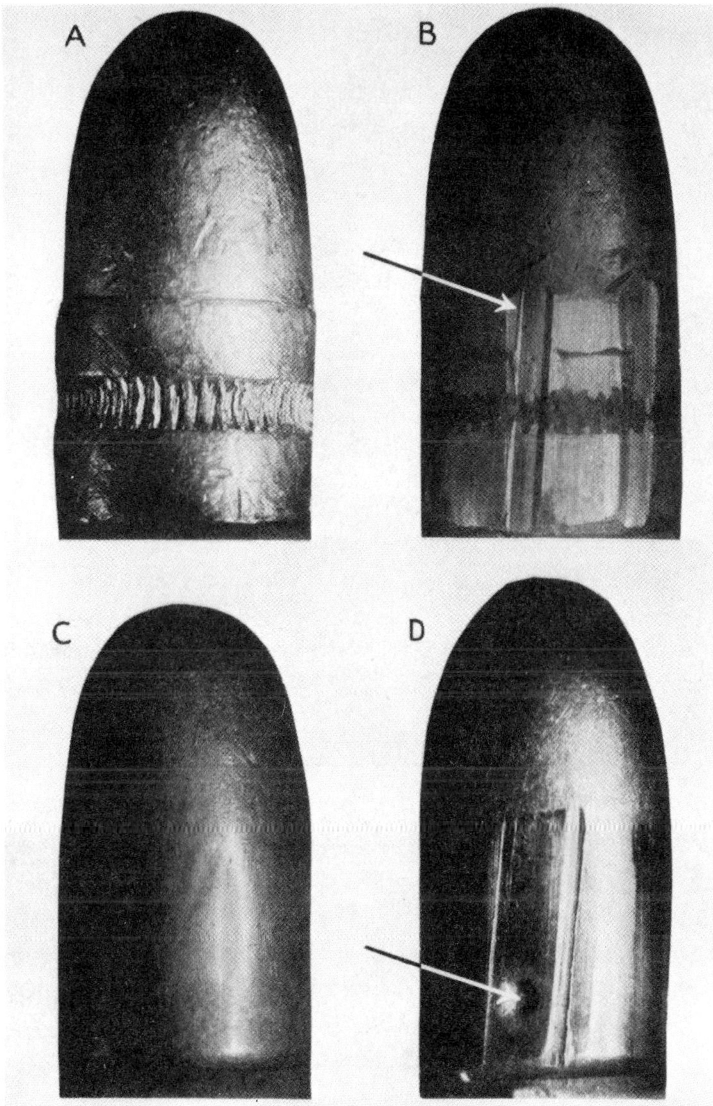

UNFIRED AND FIRED BULLETS

A is an unfired lead ·380 revolver bullet. B a similar bullet which has been fired by a Webley revolver. The Webley type of rifling (7 grooves, right-hand twist, narrow lands) is distinctly engraved on the bullet. The arrow indicates the "skid mark." C is an unfired nickel-jacketed 7·63 mm. pistol bullet. And D is a similar bullet fired by a Schwartloze self-loading pistol. The engraving is that of the type of rifling commonly used in the early models of self-loading pistols (4 grooves, right-hand twist, lands narrower than the grooves). The arrow indicates the "stabbing" mark by which the bullet is held in the cartridge case

similar to that in which the wheel of a car which is travelling just off the direction of a tram-line will skid into that tram-line and travel along it. But the mark of the first impact with the rifling remains, and constitutes what can best be described as a " Skid Mark " at the front end of the land furrow.

If the chamber is very closely bored, and the leed, or chamber cone connecting the chamber with the rifling, is short, and the bullet is a tight fit, the bullet may enter the rifling absolutely nose on. In this case there will be no skid-mark when the bullet is recovered after firing. But in revolvers in which the cylinder may not revolve so as to bring the chamber in absolute collimation with the bore ; and in cheap self-loading pistols in which the chamber may be on the loose side and the leed rather long, skid-marks are far more frequently present than absent. They mean nothing and are of no utility whatever in helping to identify an individual weapon, as they are not by any means always present on bullets fired from the same weapon. For bullets vary slightly in diameter, and a large bullet may easily escape any sort of initial wobble and enter the rifling nose on, while a small bullet may enter the rifling at a slight angle. In such circumstances the large bullet will have no skid-mark and the small bullet will have such a mark, in spite of the fact that both were fired from the same weapon.

There are various types of rifling, and these types differ widely. For example they vary in the number of grooves ; in the relative width of lands and grooves ; and in the direction of the twist, that is the twist may be right handed or left handed.

The rifling used in revolvers and self-loading pistols

may be divided conveniently into the following five types :

STEYR TYPE.—Four grooves ; right-hand twist ; grooves and lands of equal width. Used in all earlier self-loading pistols, such as the Borchardt.

BROWNING TYPE.—Six grooves ; right-hand twist ; narrow lands and broad grooves. Much the most common.

COLT TYPE.—Six grooves ; left-hand twist ; narrow lands and broad grooves. Used in all Colt revolvers and self-loading pistols, in Bayard pistols and Spanish copies of Colt pistols.

WEBLEY TYPE.—Seven grooves ; right-hand twist ; narrow lands and broad grooves. Used in all Webley revolvers.

SMITH AND WESSON TYPE.—Five grooves ; right-hand twist ; grooves and lands of equal width. Used in all Smith and Wesson revolvers except the ·45 Model 1917, Harrington and Richardson revolvers, and Iver Johnson revolvers, but in no self-loader.

In addition to the general principles adopted in these five main types there are other minor differences which may exist in rifling of the same type. For example, in the Colt type the lands in different makes of weapons may all be appreciably narrower than the grooves, but in barrels of the same calibre but of different manufacture the actual width of the lands may vary.

Similarly, some makers use deeper grooves than others, while the actual pitch, that is the angle of the twist, is usually different in different makes.

All these differences in rifling are due to definite intention in manufacture, but there are also differences which may be termed accidental. For instance the depth of groove is by no means a constant quantity in weapons which are manufactured as cheaply as possible in large numbers, and there are similar variations in the width of groove, and quite possibly in the pitch.

I propose to deal more fully with this aspect of the problem in the last chapter, but have referred to it now because it explains how it is that bullets fired through similar weapons of the same make can exhibit such marked differences in engraving, differences which I know have frequently puzzled those who have been unfamiliar with details of manufacture.

It is essential that the possibility of variations in the major characteristics of barrels—that is of depth and width of grooves, diameter of the bore, and possibly pitch of rifling—should be appreciated before any attempt can be made at identifying firearms by means of fired bullets. For frequently it is possible to say quite definitely that a particular bullet could not have been fired by a particular weapon merely by an examination of the major characteristics of the engraving, and without resorting to long and laborious examination under a microscope.

By this I do not refer to the obvious differences which can be seen with the naked eye, such as whether the twist is right hand or left hand, or whether the lands are narrower than the grooves or of about the same width. I refer rather to the *degree* of engraving, that is whether a bullet is heavily engraved all round its circumference, or whether it is only lightly touched by the lands and hardly at all, or even not at all, by the grooves. For bullets do not always "bottom" the grooves in a rifled barrel because they vary both in original diameter and in the degree to which they are "set up" by the pressure of the powder gases on the base. Lead bullets are softer than nickel-jacketed bullets, and consequently set up more readily during their journey along the bore. On this account lead bullets usually show very complete engraving all round their circumference.

But nickel-jacketed bullets do not expand so easily under the pressure of the powder gases on the base, and consequently such bullets are seldom engraved all round the circumference unless the unfired bullet happened to fit the bore very tightly. Bullets vary in diameter, just as do the bores of different weapons, and consequently a large diametered bullet will be more heavily engraved than one which is smaller in diameter if both are fired through the same barrel.

Frequently a nickel bullet only touches the bottoms of the grooves in the very middles of the grooves, and then the groove engraving shows as a small scrape mid-way between the deep furrows engraved by the lands. And occasionally a bullet fits a barrel so loosely that there is no groove engraving at all, the only marks being the furrows cut by the lands.

Nickel-jacketed bullets with different types and degrees of engraving are shown on Plate XXVI.

This variation in the degree of engraving can be of the greatest help in establishing negative evidence of identification, as will be realised from the following illustration.

If two fired bullets are to be examined with a view to ascertaining whether both have been fired from the same weapon, the first thing to do is to examine the general type of engraving, and if this appears similar the next step is to examine the character, that is whether

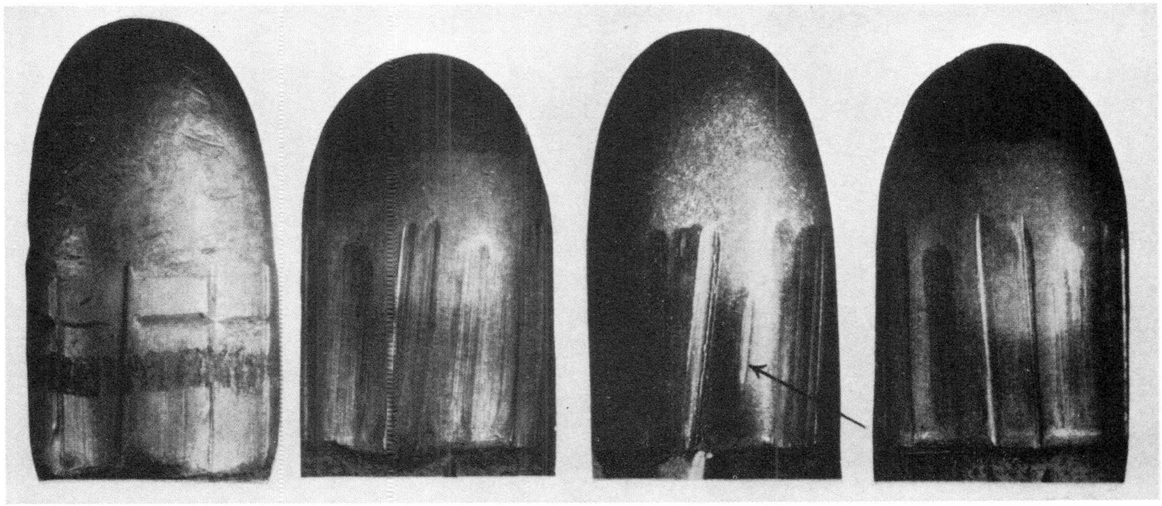

FOUR FIRED BULLETS SHOWING DIFFERENT TYPES OF RIFLING ENGRAVING

On the left is a ·380 lead revolver bullet with the Smith and Wesson type of engraving (5 grooves, right-hand twist, lands and grooves of equal width). Next is a 9 mm. nickel-jacketed bullet fired from an Astra self-loading pistol. This bullet fitted the bore tightly and is engraved all round its circumference, the striations made by the grooves being clearly visible between the deep land furrows. The rifling is the Browning type (6 grooves, right-hand twist, narrow lands and wide grooves). Next comes a 7·63 mm. nickel-jacketed bullet fired by a German Mauser pistol. This bullet only just "bottomed" the grooves in the very middle, and the groove engraving is indicated by the arrow. The rest of the space between the land furrows is free of any engraving. The rifling is also of the Browning type, but the land furrows do not appear so wide as in those on the Astra bullet because they were not cut so deeply owing to the loose fit of the bullet in the bore. On the right is a ·380 nickel-jacketed bullet fired by a Colt self-loading pistol and engraved with the Colt type of rifling (6 grooves, left-hand twist, narrow lands, wide grooves).

the engraving is heavy or light. These preliminary examinations can be carried out with any ordinary reading glass, or even the naked eye.

If one bullet is heavily engraved all round its circumference, and the other bullet is only lightly engraved by the lands and hardly at all by the grooves it can only mean that the first bullet fitted the bore of the weapon which fired it much more tightly than did the second. The next step is to measure the mean diameters of the two fired bullets, which can best be done by measuring the diameters of each bullet in five or six different directions and taking the average result. If the diameters of the bullets are the same, or that of the lightly engraved bullet is the larger, it can only mean that the two bullets could not possibly have been fired through the same barrel, in which case identity is definitely disproved.

If, however, the heavily engraved bullet is larger in diameter it is possible for both bullets to have been fired through the same barrel and further work must be conducted.

It should be understood clearly that invaluable as this investigation of what may be termed major characteristics may be, it is of utility only when the types of rifling on the two bullets to be examined are the same. Bullets are naturally less likely to bottom all across the grooves if the grooves are very deep than if they are shallow, a point which should always be kept in mind.

Then it sometimes happens that a bullet which does not fit the bore tightly does not travel down the bore in an absolutely straight line, but that it moves forwards with a very slight spiral motion. In such circumstances one side of the bullet will hug the side of the bore very tightly, while the other side will touch it but lightly. Such a bullet is engraved heavily all round one portion of the circumference while the other portion is hardly touched by the rifling at all. When engraving of this sort is encountered it provides quite definite proof that the bullet must have been a loose fit in the bore of the weapon from which it was fired, a fact which may prove of considerable help in forming an opinion as to the identity or non-identity of a particular weapon.

I would, however, emphasise again that this variation in degree of engraving is only likely to be encountered in the case of bullets with envelopes of nickel or other hard alloy. Lead bullets set up so much more readily that they are usually engraved all round the circumference; and it is, therefore, necessary to examine the minor, or secondary, engraving in their case just as it is in the case of nickel bullets in which the major engraving appears similar.

So let us now turn to a consideration of the principles governing the task of establishing the identity of a particular weapon by means of the secondary markings on fired bullets.

These secondary markings consist of a number of parallel striations in the furrows cut by the lands, and also in between these land furrows. These striations are all caused by minute tool, or other marks on the surface of the bore which scratch the outside surface of the bullet as it passes along the bore. In fact, every barrel leaves its thumb-mark on every bullet which is fired through it just as every breech face leaves its thumb-mark on the base of every fired cartridge case.

Plate XXVII shows a portion of the surface of a fired lead bullet on which the striations, or secondary markings, can be seen clearly.

But the thumb-mark of a barrel is, except in one particular, far more difficult to read than the thumb-mark of a breech face. To begin with the striations which make it up are altogether finer than any markings impressed on the base of a cartridge, and this means that a microscope is essential from the very start and that a somewhat higher power is required than is necessary for a cartridge case. Then the imprint of a barrel on a bullet is a *sliding* imprint, and is consequently far more subject to variations than is a *static* imprint such as is obtained on the base of a cartridge. In fact, one of the most surprising things which must strike any observer who is examining fired bullets is the astonishing differences which seem to be present on bullets which are known to have been fired through the same barrel. These differences are due to the sliding imprint, but with practice it is possible to detect the difference between variations resulting from the sliding imprint and variations due to different barrels.

The one particular in which the marks on bullets are more easily read than those on cartridge cases is the constancy of the direction of the markings. On a cartridge case the markings may be of almost any nature and run in any direction. But on a bullet they are all small lines, all of which run parallel to the major markings such as the land furrows. Consequently while it is frequently extremely difficult to see all the markings on the base of a cartridge simultaneously owing to the necessity of arranging the source of illumination differently to bring out different markings, with bullets it is merely necessary to focus the source of illumination at right angles to the longitudinal axis of the bullet and on that part of the surface of the bullet which is nearest to the microscope objective. The bullet is held in a special holder which enables it to be rotated about its longitudinal axis, and thus every portion of the engraved surface can be examined critically without having to consider any change in the direction of illumination.

The work of identification is conducted on the same general principles as those adopted with fired cartridge cases, and the method of procedure will probably be best explained by assuming that a fired bullet has been recovered from the body of some victim and that a pistol or revolver has been found in the possession of a suspected person. The question then arises, was the "crime" bullet fired from the suspect weapon?

The first step is to obtain two or three "test" bullets from the suspect weapon which can be used to compare with the "crime" bullet. In order to do this two or three shots should be fired into some substance from which the bullets can be recovered without injury or marking other than that due to the engraving of the rifling.

Nickel-jacketed bullets are comparatively easy to catch unharmed, as water provides an excellent medium. A large dustbin is quite a good receptacle for the water, but it should be quite 4 feet high, and in the case of some pistols which develop an exceptionally high velocity 6 feet of water may be required.

But water is useless for lead bullets, and so is sawdust. Mann in his classic work, *The Bullet's Flight from Powder to Target*, suggested *oiled* sawdust and

The lower right-hand part of the surface of a Fired Lead Revolver Bullet showing the fine Striations of Engraving

found this gave him perfect bullets. But I have not tried this myself, as I have found that cotton waste seems to be all that can be desired. My own bullet catcher consists of a wooden box 3 feet long by 1½ feet wide by 1½ feet deep. There is a hole 6 inches square in the middle of the front into which one shoots. The sides, back, bottom and lid of the box are all lined with thin steel plate so as to eliminate any possibility of a bullet piercing the box.

The box is filled with cotton waste and before firing the lid is bolted so as to make it absolutely secure.

In order to facilitate the actual work of recovering the bullets thin card screens are placed at intervals in the cotton waste so as to divide the box into compartments. After firing a shot one merely has to see which screens have been penetrated in order to locate the bullet in one particular compartment, after which it is found with surprising ease.

Certain experience is necessary in order to find out how tightly the cotton waste should be packed. If it is packed too tight a lead bullet is liable to be set up and deformed. The ideal degree of tightness will allow a bullet to reach within about 6 inches of the end of the box, and this degree will vary with the weapon and cartridge used. If one is uncertain it is a simple matter to try a few experimental shots with a similar weapon to the suspect weapon and use the same ammunition as is to be used for the test shots.

The number of test bullets which should be obtained is a debatable point. For reasons which I will give later five shots should be the absolute maximum fired in any investigation, and I think it might be better not to fire more than three. I do not regard one test bullet enough because bullets vary in their engraving, and an important component of the thumb-mark of the barrel may be so faintly engraved on one bullet as to escape notice unless its presence was being particularly sought.

Besides in preparing evidence it is essential to show that the engraving from a particular barrel is constant in its individualities, and this can only be shown if there are two or more test bullets which have been fired through that barrel.

So, on the whole, I am inclined to think that from the point of view of "bullet evidence" alone the best number of test bullets is probably three. But if "cartridge evidence" is also required it may be necessary to fire five shots, in which case all five bullets should be recovered and used for purposes of comparison.

The test bullets should all be examined under a pocket lens merely to discover which seems to be the most deeply engraved, and this bullet should be selected for a preliminary examination under a microscope.

As has already been explained the bullet is held in a special rotating holder, which merely consists of a spindle in a mount with a brass cup at one end of the spindle and a small milled head at the other. The mount is clamped to the stage of the microscope with the cup pointing inwards. The cup is filled with plasticine and the bullet is held simply by pressing its base against the plasticine. The holder is then adjusted so that the bullet is situated centrally in the field of view of the microscope. If it has been correctly attached to the holder and the spindle is revolved every part of the bullet can be examined in turn without having to move the holder.

In order to obtain critical illumination a small projection lamp is clamped to the stage and the beam of light focussed on the upper surface of the bullet. This beam should strike the bullet obliquely, as otherwise the striations will not be nearly so easily seen. Personally I always use the special projection lamps made for the purpose by Messrs. W. Watson and cannot imagine anything better or more efficient for visual work.

As with cartridges, a low power should be used at first; and the combination of a 3-inch objective and ×5 eye-piece will be found very satisfactory.

Every portion of the surface of the bullet should be brought in turn under critical illumination and studied carefully. The depth of focus—or "penetration," as it is termed in microscopy—of a low power objective is greater than that of one of a higher power, and if a higher power than 3-inch is used at first it will be found more difficult to gain an impression of the engraving as a whole since a smaller part will be in focus, and approximate focus, at the same time.

The bullet should be rotated completely and a look out kept for any peculiar or prominent striations. Such may be a very deep line, or a double line, or even two lines close together with a third a little further apart. It does not matter whether these striations occur in the land engraving or the groove engraving. The point to realise is that the first step is to detect some peculiarity in the engraving which is easily recognised.

For convenience in subsequent working I always make a little mark in the plasticine immediately at the base of the bullet exactly opposite the striations which I choose as my datum point, as this enables one to find them again without any delay.

The bullet holder can now be removed from the stage of the microscope and another put in its place holding a second test bullet. This bullet is then examined in the same way until the special striations which have been selected on the first bullet are detected.

This bullet can be removed from the holder and marked on the base with the point of a needle exactly opposite the datum point, after which the third, or additional test bullets should all be examined in the same way.

It will almost certainly be found that the striations which have been selected as the datum point on the first bullet are present on all the other bullets, although it is perfectly possible that they will vary quite considerably in appearance. In one bullet one of the striations (if more than one were selected) may be more prominent; while in a second bullet another line may be the more prominent, and in a third bullet even another line may be more clearly engraved. But the lines will all be present on every bullet, *and their positions relative to each other and the edges of the land furrows on either side will be constant.*

The striations which were selected as the datum point on the first bullet can then be regarded as the thumb-mark of the suspect barrel.

It may happen, however, that they are not found on all the test bullets, as one bullet may be abnormally lightly engraved on the side on which the striations selected should occur. Or one particular bullet may either be defaced in the bullet catcher, or that side may have been stripped instead of being properly engraved. This last happens more frequently than is imagined in the case of lead bullets fired from pitted barrels.

Such bullets should be discarded, as if the most important side is not engraved they cannot possibly be of much use.

Let us assume, however, that three satisfactory test bullets have been matched.

So far I have described the procedure suited to an ordinary microscope. Such an instrument can be used and definite results established. But when comparing bullets the Comparison Microscope comes into its own. The principle of this type of microscope was explained on page 136, but while my own personal opinion is that for cartridge cases it is less convenient than an ordinary single microscope, there can be no doubt that with bullets it is the greatest help, as it renders comparison more easy, more quick and more certain.

When using a comparison microscope the first test bullet is kept under one instrument all the time, and the other test bullets are placed under the other and compared in turn with the first bullet. So long as only one bullet is being examined the ordinary eye-pieces are retained in each microscope, which are used in the ordinary way. But when it is thought that the thumb mark of the barrel has been found again on the second, or subsequent, test bullet the two eye-pieces are removed and replaced by the special comparison eye-piece which gives its name to the instrument. When using this eye-piece the whole field of view is circular as before, but it is divided into two equal parts, the right-hand part belonging to the field of the right-hand microscope and the left-hand part belonging to that of the left-hand instrument. It is thus possible to compare the striations and their relative positions on two bullets with great accuracy, since if they have all been made by the same part of the same barrel and the bullets have been properly set in their holders, the two sets of striations will coincide exactly.

Plate XXVIII shows a Comparison Microscope with the comparison eye-piece in position. The bullets can be seen in their holders, while the two projection lamps are clamped in position on the two stages.

If a higher power than that given by a 3-inch objective is required a 2-inch objective can be used, but anything more powerful can only be necessary on very, very rare occasions. For it must be remembered that with the comparison eye-piece in position the tube length of the microscope, that is the distance between objective and eye-piece, is approximately doubled; which means that the magnification is increased to a corresponding degree.

Once the thumb-mark of the barrel has been definitely established it can be checked by rotating the two bullets until every portion of the engraved surfaces of both bullets has been compared. If the original thumb-mark is taken as the starting point in each bullet and then each bullet is revolved through 30 degrees at a time and examined again it should be found that the striations correspond in every instance. As has already been explained these striations may, in fact almost certainly will, vary in clearness, and in some cases they may be absent from one side of one bullet. But their relative positions will correspond and exact coincidence will be seen in the comparison eye-piece.

One method which I have seen advocated is to place a bullet under one instrument and keep it stationary

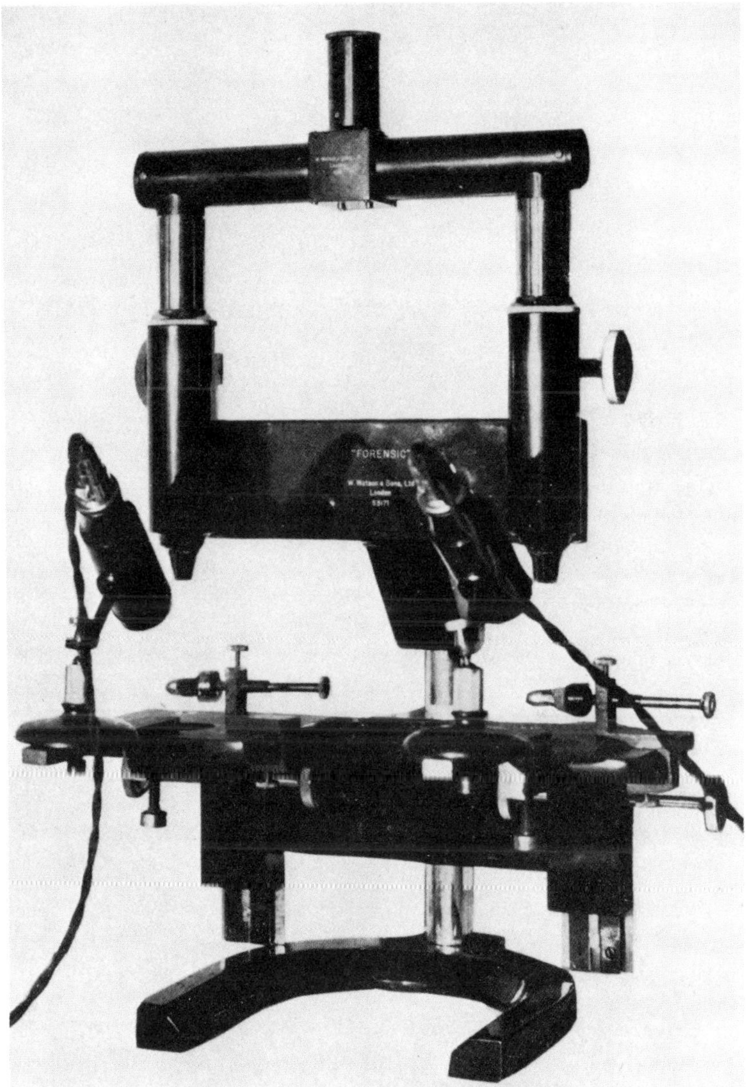

A Forensic Microscope with the Comparison Eye-piece in position

The bullets can be seen on their holders, and the two projection lamps are attached to the fronts of the stages

and revolve a second bullet under the other instrument with the comparison eye-piece in position. In course of the rotation there must come a point when the two sets of striations will coincide if both bullets have been fired through the same barrel. This method is excellent in theory but personally I have found it much more difficult to adopt successfully in practice than the method which I have suggested. If a prominent marking is selected on one bullet as a datum point there seems to be more certainty of purpose. But different methods suit different people, and what one may find difficult another may find easy, and *vice versa.*

A further advantage of the method which I have suggested would appear to be the adoption of the same definite principle both for cartridge cases and bullets, that is of selecting a prominent marking and seeing whether it is present on all other cases or bullets fired from the same weapon.

But whatever procedure is adopted there can be no getting away from the fact that the engraving on all the test bullets will exhibit certain constant peculiarities.

The next step, therefore, is to see whether the "crime" bullet also carries this same distinctive thumb-mark.

This comparison is effected in exactly the same way as in the case of two test bullets. That is, a well-engraved test bullet is kept under one microscope with the most prominent characteristic markings in focus, and the crime bullet is examined separately under the other instrument until similar markings are believed to be present. The comparison eye-piece is then placed in position and the two sets of markings compared. If they coincide the "crime" bullet may be regarded as marrying the suspect pistol, but further proof must be sought by comparing all sides of the "crime" and "test" bullets in the way that has already been described. If perfect coincidence is obtained the weapon can be regarded as having been identified.

Plates XXIX and XXX show the steps in the work of comparing two bullets. In the left-hand photograph of Plate XXIX a portion of one bullet is seen under the left-hand microscope, and in the right photograph a corresponding portion is seen under the right-hand microscope.

The ordinary eye-pieces are then replaced by the comparison eye-piece, and a part of each bullet is seen in the composite field of view, which is reproduced in Plate XXX, the bullets being slightly rotated until the two sets of striations come into perfect collimation, as can be seen in the photograph, which is at a higher magnification on account of the increased power of the comparison eye-piece.

But even if the investigator is satisfied in his own mind that he has married the "crime" bullet to the suspect weapon, more evidence is needed to satisfy a jury. Such evidence can only be provided by means of photographs taken through the microscope, and probably the best plan is to make up a series of composite photographs on the same lines as those which were suggested in the last chapter for establishing the identity of cartridge cases. Such composite photographs of bullets are shown in Plates XXXI and XXXII and in Plates XXXIII and XXXIV.

Without such photographs to support them mere assertions by some investigator, no matter how great his reputation as an expert, should be regarded with

Two different Bullets which had been fired by the same Revolver as viewed under the two microscopes of the Forensic Microscope shown in Plate XXVIII

The ordinary eye-pieces are in use and the two bullets are viewed separately one after the other

extreme caution. As I stated in the last chapter the most ridiculous claims have been put forward on behalf of the comparison microscope, and there is great danger that the mere fact of its possession may endow a witness with all sorts of imaginary skill and knowledge, at least in the eyes of the jury and public. The truth is that a comparison microscope is by no means an easy instrument to use, and that its possession no more makes an investigator a competent expert than the possession of a pair of high grade guns makes a man a good shot, or the possession of a Steinway Grand makes its owner a fine pianist. In actual practice the combination of a really experienced investigator and a powerful pocket lens is far more likely to achieve correct results, even in the case of bullets, than the combination of the most costly forensic microscope and untrained observer. And this is a truth which cannot be too widely known. If, therefore, the evidence is unsupported by photographs which tell clearly their own story, that evidence should be regarded with suspicion.

And it must always be remembered that the sliding imprint which bullets receive of the thumb-mark of a barrel is liable to record only what I have termed the "family thumb-mark" of a batch of barrels. In order to appreciate the existence and limitations of this risk it is necessary to understand how the barrels of revolvers and pistols are manufactured.

Let us first take revolver barrels.

A solid steel bar is forged and drawn which has approximately the same external dimensions of the barrel required. This bar is roughly shaped and bored down its centre, when it assumes the form of a tube, and it is this tube which is rifled. This process is carried out by a small steel cutter which is attached to the end of a rod. The rod with the cutter is inserted to the far end of the tube and the cutter is held against one side of the inside surface of the tube. The machinery is then set in motion and this pulls the cutter along the inside of the tube, its edge cutting away a thin layer of metal, thus starting one groove. At the same time the tube is slowly revolved, and this revolution imparts the twist in the rifling, the actual pitch being controlled on the machine.

In some factories the twist is obtained by rotating the cutter and not the barrel ; but the general principle is the same.

When the first cut is finished the cutter is returned to exactly the same position, and the same cut is repeated so as to deepen the grove. Altogether anything from six to ten cuts may be required to make one groove, according to the type and depth of groove required.

The second and subsequent grooves are then made in the same way.

Now although it is absolutely true that the surface of no two consecutive cuts will be *absolutely* identical, it is an equally true fact that these surfaces will have a strong family likeness one to the other. And if but one cut only would make a groove, there would be a very strong family likeness running through the surfaces of all the grooves made with the same cutter. Since, however, a number of cuts are required to complete a single groove, and since there are a number of grooves in a single barrel, it will be seen that there is ample scope for differences in the surfaces of the grooves even of barrels which have been rifled consecutively. Nevertheless the surface of the *last* groove cut in one barrel may bear a distinct family resemblance to that of the *first* groove cut in the next barrel.

In order to test this possibility Messrs. Webley & Scott, Ltd., of Birmingham, whose revolvers very rightly bear a worldwide reputation for highest quality workmanship and finish, most kindly came to my assistance. They lent me four new ·38 revolvers which had been fitted with barrels which had been rifled consecutively on the same machine and with the same cutter. I fired five bullets from each of these revolvers, and was able in each case to match all the bullets fired from each individual barrel. I found also that some of the prominent striations in one groove of No. 1 barrel corresponded exactly with similar striations in one groove of No. 2 barrel. Similarly that another groove of No. 2 barrel could be matched as regards a number of its striations with one groove of No. 3 barrel.

I was unable, however, to match any groove of No. 4 barrel with one of No. 3. Nor could I match No. 1 with Nos. 3 or 4, nor No. 2 with No. 4.

It should be noted that even when the match was effected it was not a complete match, as in each case only some of the deeper striations coincided. But these *did* coincide, and so I feel confident in my own mind that it cannot be considered safe to match two bullets unless coincidence of striations can be definitely established on at least three different sides of both bullets.

It must also be remembered that the land engraving can provide invaluable evidence which is largely independent of the rifling cutter, since the land surfaces are the remnants of the surface of the inside of the tube before it was rifled, and consequently two bullets should always be matched by the land engraving as well as by the groove engraving.

Let us now turn to pistol barrels.

The general principles of manufacture of all rifled barrels are the same, but the barrels of self-loading pistols are commonly bored and rifled in one long piece which is then cut up into lengths suitable for single weapons. In some factories the barrels are thus made in pairs, while in others as many as six pistol barrels may be cut from one long one.

Now it will be quite clear that barrels made on this principle must be in quite a different category from those which are rifled separately and consecutively as far as the possession of a family thumb-mark is concerned. For if one long barrel is first completed and then cut up into two, or more, shorter barrels, each of those shorter barrels will have had every groove cut with identical strokes of the same cutter.

Of course it may happen that some of the barrels are reversed when actually fitted to the pistols, that is the original forward end becomes the breech end, and in this case the relative positions of the tool markings will be reversed. But there is an even chance that the barrels are fitted with the tool markings running in the same direction.

In *Army Ordnance* of September–October, 1933, Major J. S. Hatcher, Ordnance Department, U.S. Army, contributed a most instructive article on this subject and stated that tests had been carried out in America to determine whether it was possible to distinguish the bullets fired from barrels made in pairs.

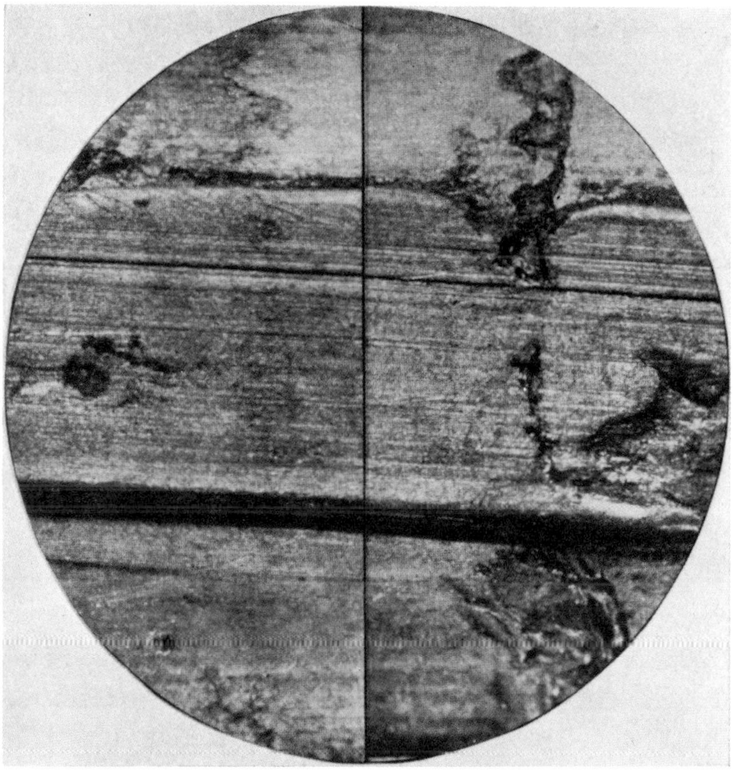

THE SAME TWO BULLETS SHOWN IN PLATE XXIX, BUT AS VIEWED WITH THE COMPARISON EYE-PIECE IN POSITION

A portion of the right-hand bullet appears in the right-hand half of the field, and a portion of the left-hand bullet in the left half. The striations on the two bullets match exactly

A rod of steel was bored and rifled under the supervision of a Government officer and then cut in two to form two barrels. This was repeated three times, making six barrels in all. Then one bullet was fired from each of these barrels and marked secretly by the officer supervising the test, who then passed them to Colonel Goddard of the Scientific Crime Detection Laboratory, Chicago. Colonel Goddard was able to marry the different bullets correctly to their respective barrels.

The explanation of this seemingly astonishing success is to be found in the blunting of the cutter during each stroke, which means that the surface of the beginning of a cut is different from that at the end of the same cut.

At the same time it must be remembered that bullets differ greatly in the clarity of their engraving, and if the bullet fits the bore rather loosely it may easily happen that it is engraved only with the family thumbmark of a particular batch of barrels.

The barrels of the very highest grade weapons are finished off after rifling with a lead lap and fine emery. Messrs. Webley always polish the barrels of their revolvers by this means, although the barrels of the four revolvers which they so kindly sent me for testing had purposely been left as they came off the machine. This final polishing undoubtedly has the effect of toning down some of the tool markings on the surfaces of both lands and grooves, and this may tend either to accentuate or reduce the existence of the family likeness. But in the great majority of self-loading pistols, and in all cheap revolvers, the barrels receive no final polishing and are left just as they come off the rifling machine.

METALLIC FOULING AND WEAR

When a number of shots are fired through a rifle small portions of the outside of the bullet are frequently removed during its passage down the bore and left adhering to the surface of the bore. At first only the smallest particles of the bullet strip, but these may become almost fused on to the inside of the barrel by the extreme heat generated, and are not easily removed. The uneven projections thus formed scrape the surface of each successive bullet and gradually increase in size until the barrel becomes badly fouled with what is termed "metallic fouling."

And since metallic fouling may easily change the nature of the engraving on a bullet it is important that the possibilities should be appreciated.

The tendency of a barrel to collect metallic fouling depends on three main factors: the smoothness or roughness of the bore; the material of which the outside of the bullet is composed; the temperature.

It is but natural that a badly pitted and rough barrel will collect metallic fouling more readily than one which is in perfect condition.

Similarly nickel possesses a far greater affinity for the steel of the barrel than lead, or other materials which are used for the envelopes of bullets.

And the higher the temperature the more readily will nickel be fused to the surface of the bore. But it should be realised that temperature can be produced in several ways: the weather; the rapidity of fire; the nature of the powder; the gas pressure; and the velocity.

The first two of these ways are obvious and so need

no further explanation, but the effect of powder and pressure are not so generally known. In Chapter IV it was explained that the gases generated by different types of powder varied in their temperature, while temperature was also seen to be a corollary of pressure.

In revolvers and pistols the powder charges are very small and the pressures low. Further, the velocities are much lower than those of modern rifles. So, other things being equal, there will never be the same tendency for revolvers or pistols to suffer from metallic fouling.

And in the case of the great majority of revolvers this tendency is still further reduced by the fact that they fire lead bullets, for lead bullets never "lead" up a barrel as nickel-jacketed bullets will "nickel" one up.

I had imagined that this fact was too well known to need emphasis, but I have since heard an expert state in court that after thirteen shots with lead bullets a barrel would be so "leaded up" as to change completely its characteristics.

Now I have again and again fired a whole box of ammunition (276 rounds) in an afternoon without cleaning my revolver until the end of the firing. And yet I have never found any lead fouling which could be noticed except at the extreme breech end of the barrel. The leading here always takes the form of a ring, or part of a ring, at the very entrance of the barrel and is due to the passage of the bullet from the chamber (in the cylinder) into the bore.

In order to check my own experience I wrote to my friends, Lieut.-Colonel A. Whitty, D.S.O., and Major H. G. Lynch-Staunton. Both these officers are former Captains of the Army Eight, and as such have trained and coached teams for the Inter-Services Revolver Match at Bisley year after year, and have thus had experience of many revolvers which is unique. Further, the experience of both officers goes back for forty years.

Colonel Whitty replied as follows :

> " I cannot say that I have ever found any real trouble from ' leading ' in the bore of the service revolver ; nor, during my many opportunities of observing others firing the ultra-rapid practices which have been so fashionable during post-War years, have I ever heard complaints from them of such trouble.
>
> " My ·455 revolver, which I got in 1902, must have had quite 15,000 rounds through it, but I have never had the bore treated for removal of lead, other than what I could do myself with the ordinary *bristle* brush. I have just this moment had a look at the bore and cannot detect the slightest sign of lead."

And Major Lynch-Staunton wrote :

> " I entirely agree with your experience of lead fouling in revolver barrels, i.e. little, if any. And when I was shooting 200 or more rounds a day I never worried about cleaning until the end of the day."

In self-loading pistols and those revolvers which can fire nickel-jacketed bullets the tendency to metallic fouling is admittedly increased, but here again there is no risk of nickelling from ten or a dozen shots fired with intervals in between so as to avoid heating the barrel unduly.

The alleged danger of metallic fouling is that the firing of more than one or two shots to obtain test bullets for comparing with a " crime " bullet may so change the surface of the bore that the striations on the final test bullet will differ from those on the first. But this danger is imaginary, and there is no risk of changing the thumb-mark of a barrel by accumulating

Two different ·32 Lead Revolver Bullets both fired by the same Revolver

These bullets are the same as those which were shown in Plates XXIX and XXX, and are reproduced again in order to show how photo-micrographs should be prepared as evidence. This Plate should be studied in conjunction with Plate XXXII

metallic fouling from firing six to ten shots through any revolver or pistol.

The effects of wear, however, are in a different category. But first of all it should be appreciated that wear can have two separate effects on the inside of a barrel. It can enlarge the diameter of the bore, and it can change the minute markings on the surface of the bore which cause the striations in the engraving on a fired bullet.

Let us first consider the possibility of enlargement of the bore.

The higher the velocity of the bullet the greater must be the wear caused by every round, and consequently if we use rifles as examples we can be safe in the knowledge that the effects of wear will be considerably greater than in any pistol or revolver, and so our conclusions will err on the side of exaggeration.

By far the best method of checking the gradual enlargement of the bore of a rifle through wear is by adopting the normal routine of the scientific target shot. A part of his equipment is a set of barrel gauges, each of which is a perfect cylinder of a certain definite diameter which is measured accurately to one quarter of one thousandth of an inch. These gauges increase in size, so that a complete set for a ·303 rifle comprises the following nominal sizes: ·303, ·30325, ·3035, ·30375, ·304, ·30425, ·3045, ·30475 and ·305.

Before shooting any series the barrel is cleaned and gauged, that is the size of the largest gauge, or plug, which it will take is noted. After the end of every series the barrel is cleaned again, and again gauged. It is then found that in course of time the barrel will take a gauge the next size larger, and this increase in the bore diameter which has brought about the change is due to wear.

Incidentally it may be stated that the size of the bore can, in some rifles, have a pronounced effect on the accuracy ; and before now I have changed an indifferent shooting barrel into a first-class barrel by polishing out the bore carefully with diamantine until it was enlarged from ·303 to ·30325.

Now I have on numerous occasions in experimental shoots fired series of 60 consecutive shots at 1,100 yards, and I have never found that the bore of any of my rifles was enlarged by one of these series sufficiently to take the next larger gauge. In fact, I would place 200 rounds as the absolute minimum number of rounds required to enlarge the bore of a ·303 rifle by ·0001 of an inch. This number could be more than doubled in the case of any ordinary self-loading pistol before a similar enlargement could be produced.

I have heard it suggested that the firing of a number of test rounds might so enlarge the bore that the bullets fired at the end of the series would no longer bottom the grooves and so present a different characteristic engraving to those fired at the beginning of the series. The only reply to this is that, in the first place, it is the land diameter of the bore which is chiefly affected by wear, and so the bullets would be more likely to bottom the grooves at the end of a long series than at the beginning ; and secondly that the series would have to consist of at least a thousand rounds before any noticeable effect could be produced.

It can, therefore, be accepted that the major characteristics of the bore cannot be changed by the wear produced by firing any ordinary number of test rounds in a suspect weapon.

But wear can, and does, have a decided effect on the minute markings on the surface of the bore which cause the finer striations in the engraving of a fired bullet.

I have long suspected the possibility of this effect, and during the past few months have carried out a number of experiments with various weapons. My experience has been that in the case of a revolver firing lead bullets, fifty rounds is sufficient to change completely the finer striations in the engraving. In some revolvers which marked their bullets with one or two very deep striations these particular striations could still be identified, although they were no longer so pronounced. But the finer striations had changed, and it was no longer possible to match the bullets perfectly with those which had been fired before the series of fifty rounds had been fired.

In the case of self-loading pistols firing nickel-jacketed bullets the change seems to be brought about more rapidly, and a series of but twenty-five rounds proved sufficient to change completely the finer striations in the engraving produced by a 7·63 mm. Mauser self-loading pistol. This pistol admittedly has a higher velocity than most self-loaders, which was my reason for using it for a test; but I feel certain in my own mind that no barrel can be relied upon to produce identical engraving after twenty rounds have been fired through it. It was for this reason that I stated in the last chapter that if bullet evidence of identification was to be utilised, not more than ten rounds in all should be fired through the barrel; five by the prosecution and five by the defence.

It must not be imagined that this type of wear will gradually leave the barrel perfectly smooth inside. As one lot of markings are rubbed away, fresh markings take their place, and so on. I have in my possession a ·455 revolver through which I have fired more than 20,000 rounds. The bore is still in perfect condition, and bullets fired from it are marked with fine striations. But I know well enough that those striations are very different from those which must have been present on bullets which were fired from it when it was new.

BULLET EVIDENCE AND CARTRIDGE EVIDENCE COMPARED

When evidence of the identity of some particular arm can be established both by the evidence of a fired bullet and that of a fired cartridge case, the question may arise as to which is the more reliable. There can only be one answer : both are equally reliable *provided both can be put forward in a manner which can be appreciated and understood by a non-technical jury.* This is the crux of the whole matter, for it is far more difficult to establish identity by means of a fired bullet than by means of a fired cartridge case. Apart from the obvious possibility of the bullet being so deformed as to render all examination abortive, there is far more chance of a good thumb-mark of any weapon being imprinted on the base of a fired cartridge case by static pressure, than it has of being engraved on a bullet by a sliding pressure. And the thumb-mark on the cartridge is not only far more easy to read, but it is far more easy to photograph than is that on a bullet.

Further, photographs of the bases of cartridge cases

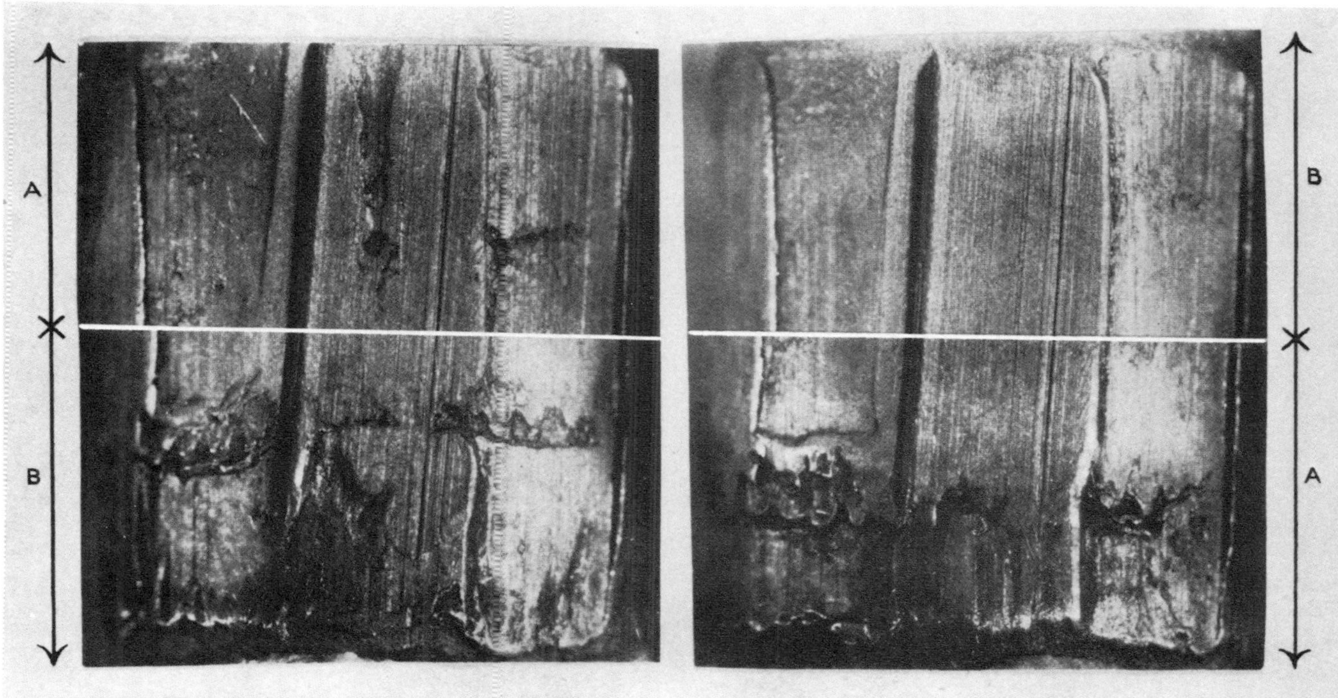

COMPOSITE PHOTOGRAPHS SHOWING HOW THE MARKINGS ON BULLETS FIRED BY THE SAME WEAPON MATCH EXACTLY

A and B are the two bullets shown in Plate XXXI. In this Plate the upper part of the left-hand photograph is A, while the lower part is B; and in the right-hand photograph the upper part is B, while the lower is A. In both photographs the striations match exactly

are more readily appreciated and understood by a non-technical jury than those of the striations on fired bullets.

In the Gutteridge case the War Office experts who did all the work of identification effected this identification both by means of a fired cartridge case and by two fired bullets. Yet the prosecution only brought forward the cartridge evidence as the bullet evidence was considered too complicated to explain satisfactorily to a jury.

There is also the chance that a sufficient number of rounds may have been fired between the firing of the " crime " bullet and the finding of the suspect pistol to render the task of identification by means of the bullet impossible. But the breech face of a weapon is unaffected by wear, although it may in time become pitted by corrosion. But such a process is slow and will probably take years, and even then some of the major markings may remain sufficiently distinct to enable identity to be established.

Naturally the idea must present itself of the possibility of a criminal altering the thumb-mark of his weapon. It would be a comparatively easy matter to alter the thumb-mark of a barrel, and this fact must always constitute a possible weakness in the work of identification by means of bullets. At the same time if the barrel were scoured out crudely, as it probably would be, the results should be quite obvious to any competent firearms expert, and the very fact that the barrel had been so treated would cast grave suspicion on the owner of the weapon.

But a breech face is far more difficult to alter. In fact, in the case of many makes of self-loading pistols it would require a competent worker to strip the mechanism so as to be able to get at the breech face with a tool. The breech face of a revolver would easily be filed. But such filing would again be obvious and merely throw suspicion on the owner of the revolver.

CHAPTER VIII

THE IDENTIFICATION OF FIREARMS. TECHNIQUE AND EXAMPLES

IN the last two chapters it has been seen that the science of identifying individual firearms is dependent on the microscope. Now a microscope is by no means an easy instrument to use. There is a common belief that almost anyone can use a microscope with effect; but like many common beliefs it is quite erroneous. In order to interpret correctly the result which the power of the microscope can convey it is essential that the investigator should master the correct technique. There are various excellent and instructive books on the subject, so any attempt to explain here even the elementary principles would be entirely superfluous. But there are three cardinal principles which are absolutely vital to success, and which cannot be emphasised too strongly nor too often. These principles are:

(1) Never use a higher power than is absolutely necessary.

(2) Correct illumination of the object is more important than anything else.

(3) Constant practice trains the eye to see detail which would escape an untrained eye.

Let us take these principles in turn.

POWER.—The rule that it is a fatal mistake to use

a higher power than is absolutely necessary should be familiar to all who have ever even tried to use a microscope. It is elementary, and known to every student when he first begins work. For the higher the power the greater the attendant difficulties. If detail can be distinguished with a particular power it is less easily distinguished if the power is increased.

And quite apart from this an increase in power means a corresponding increase in the difficulties of examining either fired cartridge cases or fired bullets. It must always be kept in mind that identification is not established by one single mark, but by the relative positions of a number of marks; and consequently the fewer of these marks which can be seen in the field of view simultaneously the greater the difficulty of assessing their value. For it should be quite obvious that an increase in power means a corresponding decrease on the field of view. So if an unnecessarily high power is used one cannot see the wood for the leaves on the trees. It is really like trying to ascertain the relative positions on the Thames of Tilbury and Maidenhead by studying a large-scale plan of Waterloo Bridge.

Then there is another disadvantage of an increase in power, namely loss in penetration, or depth of focus. On the curved surface of a bullet, in particular, only one small strip of the surface can be brought into focus at a time if, say, a 1½-inch objective is used; and this prevents one from noting the relative positions of the striations. But if a 3-inch objective is used the greater part of the bullet can be seen in focus and an accurate idea formed as to the relative positions of the striations.

For cartridge cases the most generally useful objective

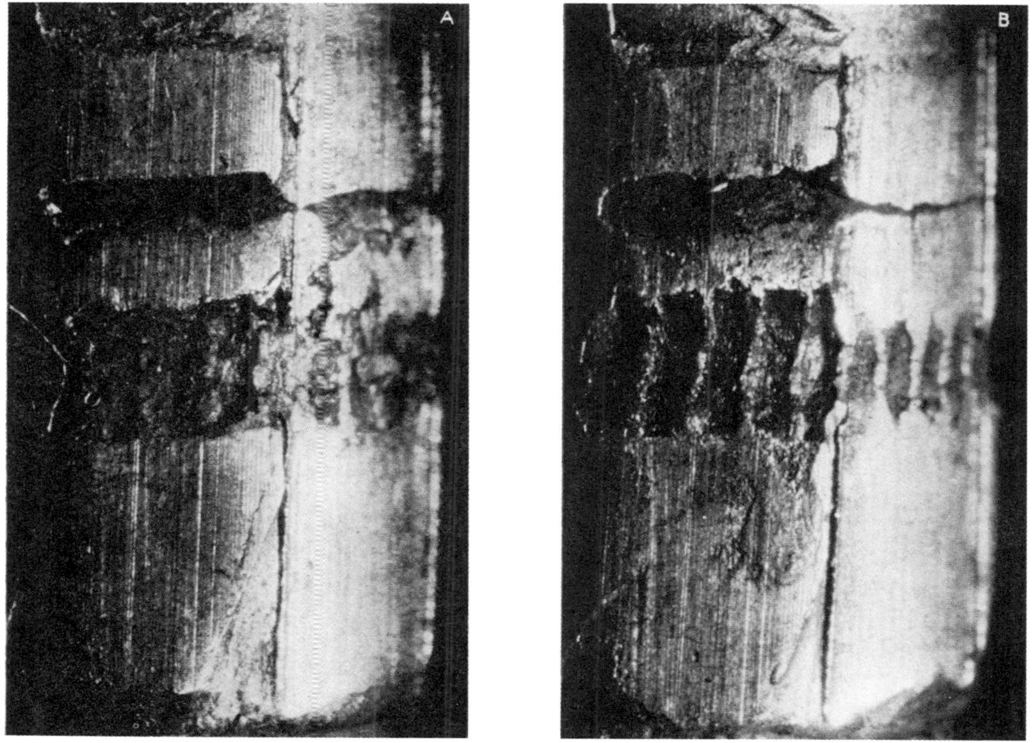

Two different ·38 Lead Revolver Bullets both fired by the same Revolver

The striations on these bullets are finer than those on the bullets shown in Plates XXIX to XXXII, and consequently a higher magnification has been employed which renders it only possible to show a portion of the bullets in the Plate

is a 3-inch, and for bullets a 3-inch and 2-inch. No more powerful objective should ever be required. My own battery of objectives ranges from a 4-inch to a $\frac{1}{12}$-inch Oil Immersion. But after numerous trials I have come to the conclusion that any more powerful objective than a 2-inch tends to confuse rather than help when examining fired cartridge cases or bullets.

I have found that the most generally useful ocular, or eye-piece, is a $\times 5$. If an investigator is very anxious to try a higher power than the combination of this ocular and a 2-inch objective he can either use a slightly more powerful ocular or even increase his tube length by racking or pulling out the draw-tube. Either method will reduce the field of view, but the penetration will not be materially affected. Of course, when using high-power objectives, particularly apochromats, it is essential to use the correct tube length for the objective in use. But low power objcctives are not nearly so sensitive to changes in tube length and permit liberties in this direction which would be impossible with high powers.

ILLUMINATION.—It is impossible to overestimate the vital importance of illumination. It must be realised that it is only possible to see the finer striations on both fired bullets and fired cartridge cases by seeing their shadows, and consequently the light source must be placed so as to fall obliquely on the surface under examination. Also that any striation will be seen more clearly if the light strikes it at right angles. If the direction of the light coincides with that of the striation, that striation will be quite invisible.

From this it is a corollary that correct illumination

is more important as regards rendering detail visible than power.

It will also be obvious that if two fired cartridge cases or two fired bullets are to be compared the source of light must strike them both at identical angles. If, therefore, photographs of two cartridge cases or two bullets are ever brought forward for purposes of comparison the first point to look at is the direction of the illumination in each case. If the direction is not the same the photographs do not deserve any consideration whatever.

Practice.—It is astonishing what practice can do in the way of training the eye to detect markings. For this reason anyone who is interested in the subject should make a point of examining as many fired cartridge cases and bullets as he possibly can, and keep on examining them. Naturally some men will show greater aptitude than others, but all will be helped by constant practice.

And here it may be mentioned that a Greenough stereoscopic microscope can at times be useful for showing up faint marks, especially if the illumination is not absolutely correct.

THE FORENSIC MICROSCOPE

The comparison microscope shown in Plate XXXVIII was made specially for me by Messrs. W. Watson & Sons, and at my request it was kept as simple as possible. The result is that various movements not usually found on ordinary microscopes are included, while other movements have been deliberately omitted for the sake of simplicity. For example, there is no fine adjustment, as such is quite unnecessary for the low-power objectives which are used. On the other hand, there is both body and stage focussing, and an altogether exceptional range of focus. Further, the stages can be removed completely from the stand, thus allowing one to examine the breech face of some weapon with a freedom which would be impossible were a stage in position.

A mechanical and revolving stage is a great convenience when examining cartridge cases, but of no practical help when examining bullets. And since I keep this instrument for bullets and use another microscope for cartridge cases plain stages were fitted for reasons of simplicity and expense.

Many observers prefer using a microscope in the vertical position. Personally I find the horizontal position more comfortable and less tiring, but much depends on custom and usage. However, by reversing the stages and limb on the stand, and using the stand as a tripod instead of with thc horseshoe foot on the ground, an instrument of the same type as my own can be adapted easily to an inclined position, which many observers will find more comfortable to use.

The instrument set up in this position is shown on Plate XXXV.

PHOTO-MICROGRAPHY

As has already been stated any evidence of identification which is unsupported by photographs cannot be regarded as being anything more than an expression of opinion. Photographs are, accordingly, essential; and such as are deemed necessary must be taken through the microscope.

Once again it must be emphasised that no higher power than is absolutely necessary to show the detail should ever be used. And in the case of photographs which are to be handed to a non-technical jury this rule is more important than ever. For if a high-power photograph of, say, the engraving of the land on a fired bullet is shown, the majority of the jury will probably not understand what they are looking at. But if the greater part of the bullet is included in the picture they will follow the argument for, or against, identification much more readily.

And illumination is even more important than in visual work.

The best source of illumination is certainly an arc-lamp. A Pointolite is also excellent, but expensive. Probably an ordinary 12-volt 36-watt bulb for a motor headlight is as good as anything, especially when fitted in a special projection lamp made for the purpose by Messrs. W. Watson.

It is always better to work on a light of approximately one wave length, as this helps to bring out the detail and increase contrast. Personally I have found Wratten K2 and K3 filters excellent; also filters of signal green and apple green.

Ordinary microscope objectives can be used, but since an eye-piece will not be required for the low powers which are used—an eye-piece need not be used for any magnification much below 50 diameters—the special photo-micrographic objectives such as are used for macro work in metallography are certainly better in every way.

The greatest difficulty will probably be vibration. It is astonishing how sensitive the combination of

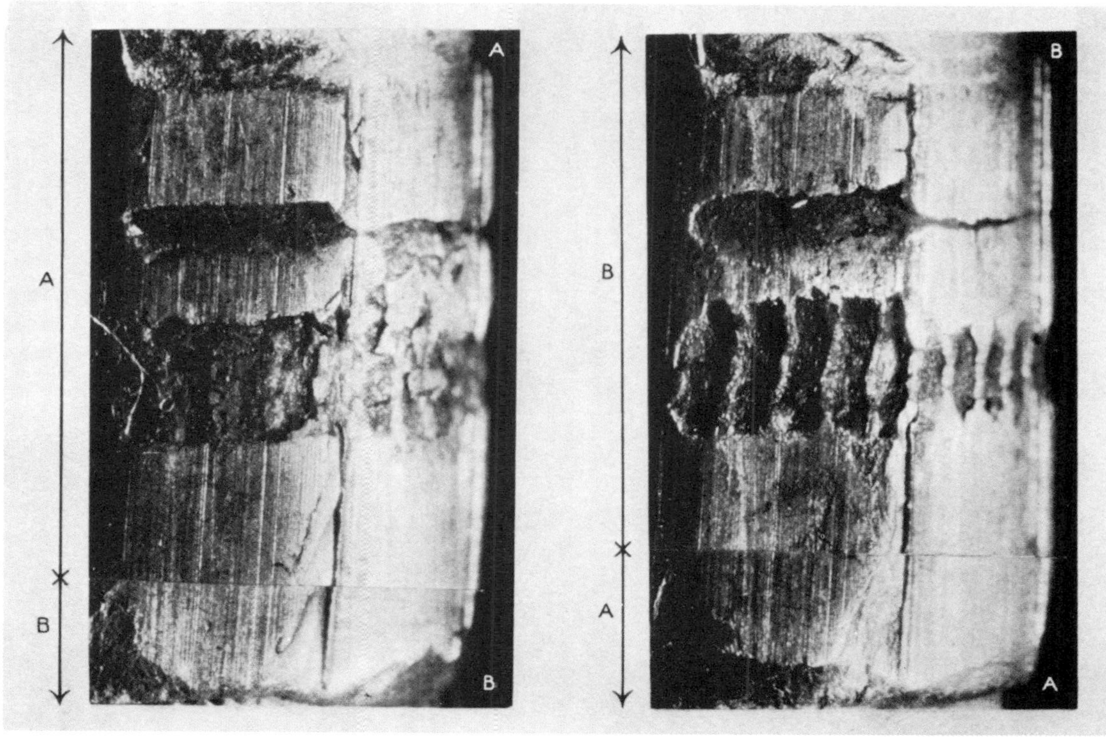

COMPOSITE PHOTOGRAPHS SHOWING HOW THE STRIATIONS ON BULLETS FIRED BY THE SAME WEAPON MATCH EXACTLY

A and B are the two bullets shown in Plate XXXIII. In this Plate the upper part of the left-hand photograph is A, while the lower part is B; and in the right-hand photograph the upper part is B, while the lower is A. In both photographs the striations match exactly

microscope and camera can be to vibration, and all sorts of devices have been invented for overcoming this difficulty. For the slightest sign of vibration will be enough to blur the photograph. For this reason I prefer personally a horizontal camera. Vertical cameras are usually clamped to a heavy iron bar which stands up from a solid base. Massive though such a bar may be, the very fact that it is fixed at its bottom end and free at the top renders it peculiarly susceptible to vibration and it has a tendency to vibrate like a great tuning fork on what may appear to be ridiculously slight provocation.

And quite apart from vibration I find myself that a horizontal camera is more convenient to use. But such points must always be a matter of personal opinion.

DISTORTION IN PHOTOGRAPHY

In view of the fact that photography must always play a most important rôle in the evidence of identification of firearms certain limitations which it possesses should be understood. The most important of these is Distortion.

Probably every amateur has at some time or another taken a photograph of some relative or friend in a more or less recumbent position with the feet much nearer the camera than the head. The result is not usually flattering since in the photograph the feet may quite likely seem to be unnecessarily large even for a real son of Anak, let alone an ordinary human.

This exaggeration of the size of objects which are relatively near to the camera is the commonest form of photographic distortion, and the shorter the focal length of the lens used the greater the exaggeration.

For this reason it is impossible to obtain any accurate comparison of the size or position of two objects by means of photographs. But this point is best illustrated by an actual example.

Plate XXXVI shows two photographs of the gun lock which is also shown in Plates X and XI, but in those photographs it is in position on the gun.

These two photographs in Plate XXXVI have both been taken from exactly the same angle, but in the upper photograph a long-focus lens was used, while in the lower photograph a short-focus lens was used. The result is that in the upper photograph it is possible to see daylight between the curve of the hammer and the top edge of the lock plate, while the protrusion of the sear tail below the bottom edge of the lock plate (see arrow) is pronounced. But in the lower photograph, there is no daylight between the hammer and top edge of the lock plate, while the protrusion of the sear tail is very slight.

The explanation is that with a short-focus lens it is necessary to take the photograph from a much shorter distance away in order to obtain an image of the same size on the plate. And since the hammer is nearer the lens than the lock plate, and the lock plate is nearer than the sear tail, the size of the hammer is exaggerated in comparison to that of the lock plate ; and the size of the lock plate is exaggerated in comparison to that of the sear tail ; in exactly the same way as the amateur's photograph exaggerated the feet of his victim in comparison to his head.

The result is that the hammer covers the top edge

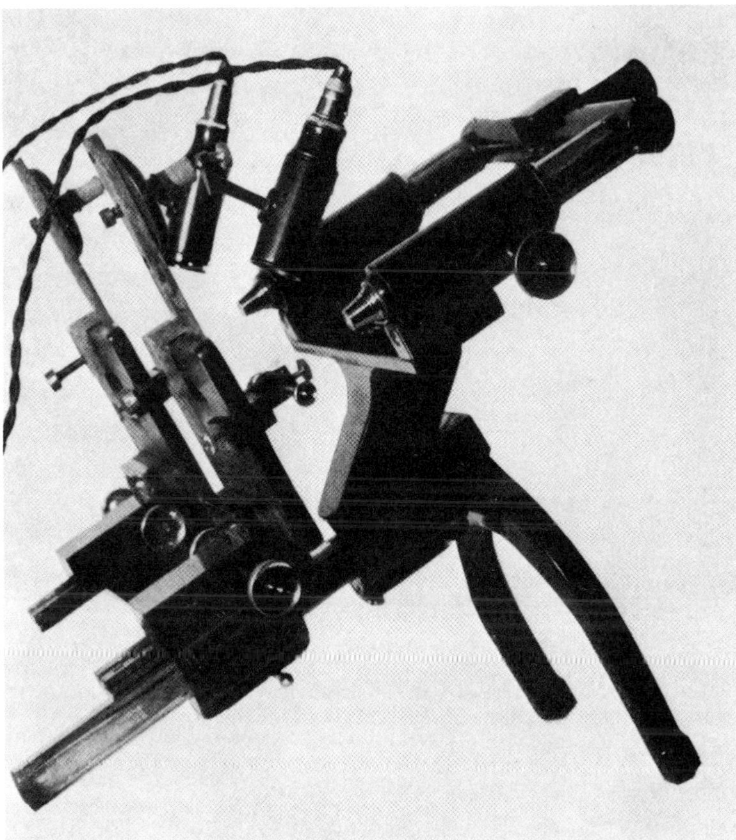

A Forensic Microscope by Watson set up with the stand reversed so as to adapt the instrument to an inclined position

of the lock plate, and the bottom edge of the lock plate covers the sear tail.

This example proves how ridiculous it would be to form any estimation of, for example, the degree of protrusion of the sear tail below the edge of the lock plate. All that photography can prove is that the sear tail *does* protrude below the bottom edge of the lock plate.

Distortion depends on the relative distance of the different parts of the object from the lens of the camera, and the focal length of the lens of the camera. If the object to be photographed could be so arranged that every portion of it was situated on the inside surface of a sphere of which the centre was the lens of the camera there would be no distortion. But failing such an impossible arrangement distortion is inevitable, and must be accepted as a fact.

Consequently photography should never be used as evidence of relative size, but only as evidence of existence, just as in the example I have given it could only be used as evidence of protrusion of the sear tail and not as evidence that the sear tail protruded by any definite amount.

For this reason I must confess that I have never been entirely happy in my own mind about the employment of composite photographs of fired bullets and cartridge cases as evidence of identity. Theoretically the principle is wrong. But in actual practice whatever error there may be is constant in both photographs, and so is of no moment. For in order to make a composite photograph the two component photographs must both be taken with the same microscope objective, and with exactly the same distance between objective and plate. All focussing must be effected by the stage, that is by moving the object. Consequently whatever tendency to distortion there may be in the resulting photographs will be the same in both.

If, however, focussing is effected by racking the objective in or out, the ratio of the distances between bullet and objective and objective and plate will not be the same in both photographs, and in that case the tendency to distortion will also differ in the two photographs. But in such circumstances it will not be possible to match the striations in the two bullets and make a perfect composite photograph, so there is a sort of automatic check on distortion.

Since composite photographs can offer most convincing evidence, and since a perfect composite photograph cannot be made up unless there is an absolute match in the two bullets or cartridge cases, I think that their use is correct and legitimate, especially as a perfect result is not likely to be achieved without knowledge and technique. But at the same time their limitations must be realised, and they should *never* be used as a basis of measurements.

For example, let us suppose that two striations on a fired bullet in the very middle of a photograph are exactly one-tenth of an inch apart. Also that two striations considerably nearer the edge of the same photograph are also one tenth of an inch apart. It would, then, be quite wrong to assume that the actual distances between these two pairs of striations was the same on the bullet. For the surface of the bullet is curved, being more or less cylindrical and on this account the distances between a number of striations, all of which are exactly the same distance apart, will

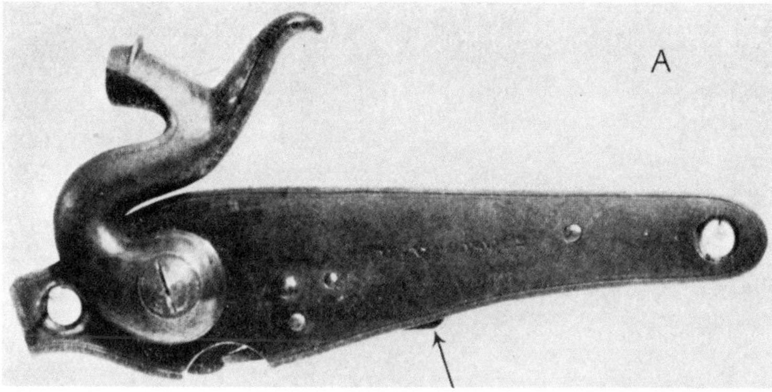

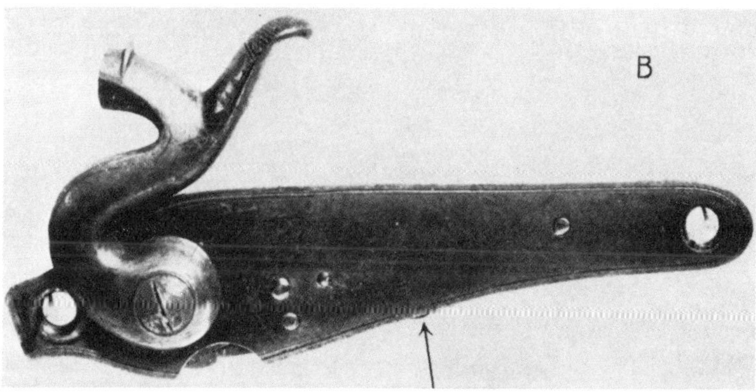

TWO PHOTOGRAPHS OF THE SAME LOCK TAKEN FROM EXACTLY THE SAME ANGLE, BUT WITH LENSES OF DIFFERENT FOCAL LENGTHS

The photograph marked A was taken with a 16-inch lens, and that marked B with a 6-inch lens. In A a streak of light can be seen between the lower curve of the hammer and the top edge of the lock plate, while in B the hammer overlaps the lock plate. Again, in A the protrusion of the sear tail below the bottom edge of the lock plate appears to be far greater than it does in B (see arrows). These two photographs emphasise the possibility of distortion in photography, particularly when lenses of different focal lengths are employed

seem to be greatest in the middle of the photograph, and smaller and smaller as the edge of the photograph is approached, and the variation in this distance will depend on the objective.

EXAMINING THE SURFACES OF BORES

An investigator will naturally wish to examine the inside of the barrel of any weapon which is passed to him with a view to identification, just as we will wish to examine the breech face. The magnification need not be high, in fact too high a magnification renders visibility much more difficult owing to the limited amount of light usually available, and the simpler the instrument the better. Messrs. J. H. Steward, Ltd., make an extremely practical and efficient instrument with which it is possible to detect any signs of metallic, or other fouling, corrosion, pitting, etc.

A good double convex lens of rather a long focus is even simpler, and very nearly as efficient. Some years ago Messrs. Steward mounted three such lenses of different foci for me in a single folding mount. This combination takes up little room in a waistcoat pocket and enables one to examine any part of the bore of the longest rifle sufficiently critically to detect metallic, or other fouling, as well as corrosion, etc. But it is not quite so satisfactory in the case of a long rifle barrel as the "Barrelscope," although for pistol and revolver barrels it is nearly as good.

If it is desired, for any particular reason, to examine the surface of the bore at a higher magnification a cystascope is the best instrument to use.

SOME ACTUAL EXAMPLES

The whole science of identifying individual weapons by means of microscopic examinations of fired bullets and fired cartridge cases has probably been studied in America longer than in any other country. But it is a comparatively new science there, and its investigation on scientific lines is really due to the appalling blunders of the prosecution in a particular murder charge.

In *Army Ordnance* of July–August, 1933, Major J. S. Hatcher stated that the present science of firearms identification in America really owes its inception to the late Mr. C. E. Waite, formerly an operative in the Department of Justice, and since his death in November, 1926, the work has been carried on by Colonel Calvin Goddard and others.

Mr. Waite first became interested in the subject in 1915 when a tenant farmer named Stielow was tried for the murder of a woman and her employer. An "expert" for the prosecution testified that under the microscope he had found nine abnormal defects in the flare of the muzzle of Stielow's revolver and had also found nine corresponding peculiar scratches on the four bullets taken from the bodies.

Cross-examination brought out that the enlarged photographs of the bullets shown to the jury by this "expert" did not reproduce the nine marks, as for some unknown reason photographs had been taken of the opposite sides of the bullets. When asked why the uneven ridges at the very extremity of the barrel should mark the bullet, the "expert" explained that the cylinder of the revolver fitted so tightly against the breech end of the barrel that there was no leakage of gas, and so the full force of the gas followed the bullet out of the muzzle. This expanded the lead as it left the muzzle and so filled in any depressions which existed at the outer edge of the bore with the result that the bullet received scratches from the elevations which existed between the depressions mentioned.

On the basis of this "expert" testimony that the bullets had been fired from the accused's revolver and could have been fired from no other, the jury found Stielow guilty.

But his lawyer began a series of appeals and firing tests were made with Stielow's revolver by competent investigators. The first test consisted of placing a piece of paper over the revolver when firing. Instead of their being no leakage of gas between cylinder and barrel the paper was set on fire.

Next fired bullets were recovered and compared under a microscope with the bullets taken from the murdered man and woman. No trace of the scratches which had doomed Stielow could be found. It was, however, discovered that one of the lands on each of the "crime" bullets was abnormal, equalling the combined width of two normal lands; while there were five lands and five grooves, all of normal width on the test bullets from Stielow's revolver.

The murders had undoubtedly been done with a defective weapon in which the rifling tool had presumably been broken, or the operator had failed to complete his operation and the error had escaped the inspector's notice.

The distinction between the two sets of bullets was glaring, and there was no possible chance that the fatal bullets could have been fired by Stielow's revolver.

THE GUTTERIDGE CASE

In England evidence based on the identification of firearms has only twice been brought forward in murder trials. The first of these was the trial of Brown and Kennedy for the murder of P.C. Gutteridge. As has already been explained the police were convinced that Brown and Kennedy were the murderers, but the final link in the chain of evidence was missing. Two revolvers were found in Brown's possession, and bullets were recovered from Gutteridge's body and a fired revolver cartridge case was found in a car known to have been used by the murderers.

The Home Office very wisely sought the help of the War Office, and experts from Woolwich Arsenal and the Royal Small Arms Factory, Enfield, proved beyond any possible doubt that one of Brown's revolvers had definitely fired the cartridge found in the car, and also that both revolvers had been used for firing the bullets recovered from Gutteridge's body.

And here it may be mentioned that these War Office experts effected the task of identifying these revolvers by means of ordinary microscopes and not by a comparison microscope. This shows what can be done by competent experts and emphasises the fact, which should be obvious, that it is not the instrument which effects the identification but the man who uses the instrument.

The manner in which the work of identification was conducted and in which the evidence was prepared was a model of how such work *should* be carried out and such evidence *should* be prepared.

One by one the marks on the breech face of the

PLATE XXXVII

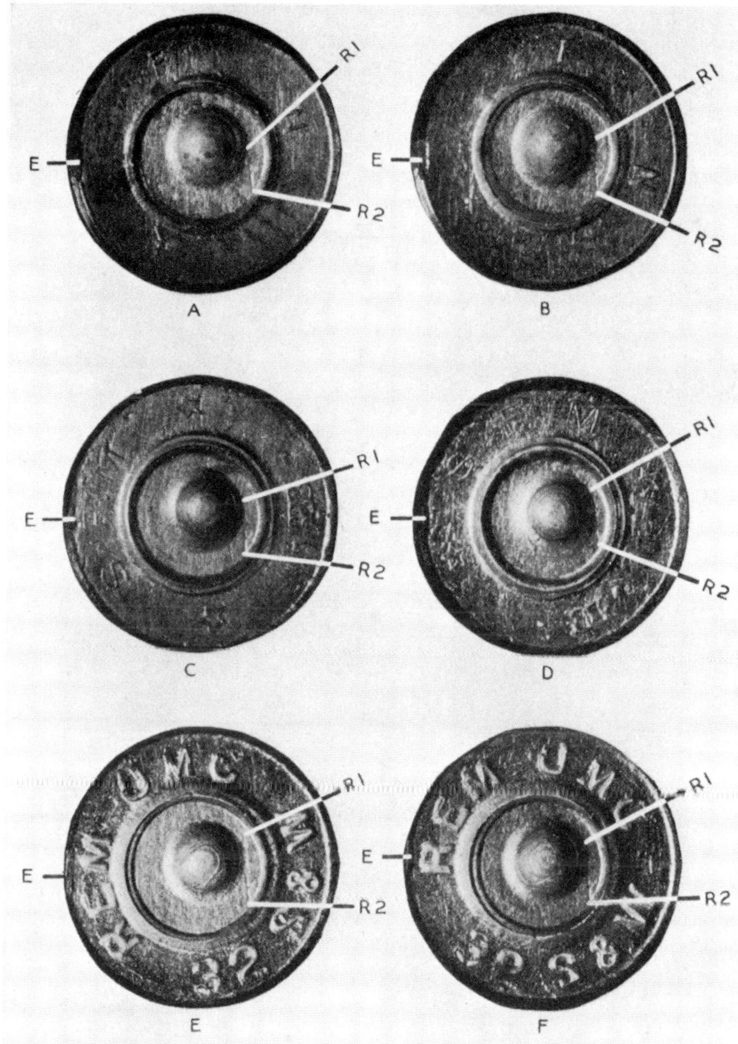

THE CASE OF THE SHOT CYPRIOT DOCTOR

Six different cartridges all fired by the pistol which was found hidden on the premises of the accused man. In each cartridge the ejector mark is a deep nick and is indicated by E, while across the cap, and opposite the ejector mark, is a pronounced ridge indicated by the lines R1 and R2. This ridge is the mark impressed by a deep tool cut which was found running across the breech face of the pistol immediately opposite the position of the ejector block. Another peculiarity constant to all six cartridges is the "ex-centricity" of the striker indentation, which is towards one o'clock

revolver were carefully noted and compared with marks on the base of the "crime" cartridge, when in every instance the two sets of markings were found to correspond.

These markings were as follows:

(1) There were two concave and overlapping indentations on the face of the shield of the revolver, one above the other, at 12 o'clock; and one indentation at 7 o'clock at the edge of the striker hole.

There were two comparable embossings on the cap of the "crime" cartridge, but the indentation on the edge of the striker hole was only partially reproduced on the cap owing to the protruding hammer nose (or striker) preventing a complete "set back" of the cap when the hammer was down in the fired position.

(2) There was a burr round the edge of the striker hole extending from 9 o'clock to 11 o'clock. This was also reproduced on the cap.

(3) The shape of the hammer nose (or striker) was irregular, the left side being inclined and the right side vertical. These irregularities were reproduced in the cap indentation of the "crime" cartridge.

(4) The extremity of the hammer nose was damaged in a peculiar manner by numerous small indentations. There were comparable embossings in the bottom of the striker indentation of the "crime" cartridge.

Photographs were taken of five "test" cartridges fired from Brown's revolver and another of the "crime" cartridge. The identity of all six cartridges was obvious, even to an untrained person.

But not satisfied with this positive evidence of identification the experts examined the whole stock of old revolvers of exactly similar pattern and manufacture which had been received from various units in the Army for repair. This stock consisted of *1,375 revolvers*, out of which six were found with irregular indentations on the breech faces. Test cartridges were fired from these six revolvers and compared with the "crime" cartridge; but in every case the markings were distinctly different from those on the "crime" cartridge.

The bullet evidence was equally complete, but owing to the extraordinary simple and convincing nature of the cartridge evidence, and the difficulty of explaining the bullet evidence to a non-technical jury, the prosecution decided not to bring the bullet evidence forward and relied entirely on the cartridge evidence.

The similarity of the photographs of the six cartridges was so pronounced that when the judge looked at them at first he thought that all six were "test" cartridges, and he asked for a photograph of the "crime" cartridge, when he was told that he already had it.

The jury found both prisoners guilty.

THE CASE OF THE SHOT CYPRIOT DOCTOR

The only other occasion on which evidence of firearms identification has been put forward in England in a murder trial was in March, 1933, and this case may be of interest as expert evidence was brought forward both by the prosecution and the defence.

The facts leading up to the trial were briefly as follows.

A man named Zemenides, who styled himself as "Doctor," was the apparent head of the Cypriot community in London.

He had, not to put too fine a point on it, swindled

a young pastry cook named Petrou, who had only recently arrived from Cyprus and could speak hardly any English, out of a sum of £9.

At 11.20 p.m. on the night of January 2nd a man knocked on the door of the house where Zemenides lived, which was opened by a Mr. Deby, a fellow-lodger. The stranger was a dark man who asked for the doctor. A few minutes later Deby heard calls for help ; and, on rushing out of his room, found Zemenides struggling with the stranger, who held a pistol in his hand. Deby shouted to the stranger not to be a fool, when the doctor took cover behind Deby. The stranger had a shot at the doctor, but missed. The doctor then tried to rush into his own room from the passage in which the struggle was taking place, but as he entered his room the stranger fired again and shot him through the heart. The stranger then bolted.

The passage in which the struggle had taken place was but dimly lit and Deby never got a chance of seeing the stranger's face clearly.

The police were later informed by a Cypriot that Petrou had used threatening language against the doctor, and that just before Christmas Petrou had bought a self-loading pistol from another Cypriot together with *seven* rounds of ammunition.

Petrou was identified by Deby and another man who had seen a stranger leave the doctor's house for the dark street. Neither of these identifications, however, were very positive, although they *were* identifications ; but they broke down under subsequent cross-examination.

Petrou was then arrested and then another Cypriot told the police that Petrou had actually confessed to him that he had shot the doctor. And a few days later the police were informed that if they searched in the cellars of the house where Petrou lodged they might find something interesting. They did so and found, hidden behind some lumber, a ·32 self-loading Browning pistol with *five* cartridges in the magazine.

Two of these cartridges were ordinary self-loading pistol cartridges, that is rimless with nickel-jacketed bullets, and three were ·32 revolver cartridges with the rims filed down and with lead bullets.

And in the passage where the doctor was shot the police found two fired cartridge cases, one of which was a rimless self-loading pistol cartridge case, and the other a ·32 revolver cartridge case with the rim filed down. Further, the bullet recovered from the doctor's body was a nickel-jacketed pistol bullet, while a lead revolver bullet was cut from the wainscoting of the passage.

The case against Petrou looked as complete as any case could be, but as a final touch, the prosecution called in the Home Office firearms expert to examine the "crime" bullets and cartridge cases. This expert did so and gave evidence at the Police Court that the bullet which had killed the doctor, and the two fired cartridge cases found in the passage, must beyond doubt have been fired by the pistol which was found hidden in Petrou's lodgings. The bullet extracted from the wainscoting was too deformed for useful examination.

It was at this juncture that I was called in by the defence, as was Mr. R. K. Wilson, whose exceptional knowledge of pistols proved quite invaluable.

I will admit at once that it never entered my head that there could be any mistake, as the evidence seemed overwhelming. But naturally I agreed to examine the "crime" bullet and cartridges in order to see whether they really did marry the suspect pistol.

In view of the fact that there were two fired cartridge cases and only one bullet, and further in view of the comparative simplicity of cartridge evidence as opposed to bullet evidence, I decided to go nap on the cartridge cases and only to use the bullet as confirmatory evidence. Mr. Wilson agreed with this plan.

Accordingly I determined to get a large number of cartridge cases all of which had been fired by the suspect pistol, and so find the thumb-mark of the pistol beyond any possible doubt. Mr. Wilson fired 50 shots into water, using self-loading pistol ammunition made by the F.N. factory, as this factory also makes the Browning pistols. I collected every one of the 50 fired nickel-jacketed bullets and 49 of the fired cartridge cases, one being lost. I then took these exhibits home and began work on the cartridge cases.

The thumb-mark of the pistol was soon obvious. On every cartridge the ejector mark was a deep nick in the rim, and running across the cap, tangential to the striker indentation and immediately opposite the ejector mark, was a pronounced "ridge," which was obviously the impression of a deep tool cut on the breech face of the pistol. This ridge and the ejector mark were so clearly marked on every cartridge that they could easily be seen with a pocket lens, while in many cases they were visible to the naked eye.

Another constant peculiarity was the "ex-centricity" of the striker indentation. When the cartridges were orientated with the ejector mark at 9 o'clock the "ex-centricity" of the striker indentation was quite constant at 1 o'clock.

I then examined the bullets and found that the bore of the pistol was obviously of very large diameter. In every case the land engraving was comparatively faint and only reached just above the cannelure of the bullet, while in not one single bullet was there groove engraving all round the circumference. The majority of the bullets had clearly never bottomed the grooves at all, as they were entirely innocent of any groove engraving; and those which did show some groove engraving were only engraved thus on one side, proving that the bullets could not have fitted the bore tightly all round.

A day or two later the police brought the "crime" bullets and fired cartridge cases to my London office, where Mr. Wilson and I examined them together and compared them with our test bullets and cartridge cases.

Until I made this examination I felt quite confident that I would find the same thumb-mark on the two "crime" cartridge cases as I had found on the 49 test cases, and which I have already described. But to my utter astonishment I found that the markings on these two "crime" cases were totally and utterly different from those on any of my test cases.

There was no sign of any ejector mark; nor was there any sign of the "ridge" across the cap. But in each striker indentation there was a slight, but perfectly distinct, striker scrape on one side. This scrape enabled me to orientate the cases, and when the scrape was placed at 9 o'clock, when the cases would be orientated similarly to the test cases with the ejector mark at 9 o'clock, the "ex-centricity" of both striker indentations was markedly at 6 o'clock.

To say that I was astounded is to put it mildly. On

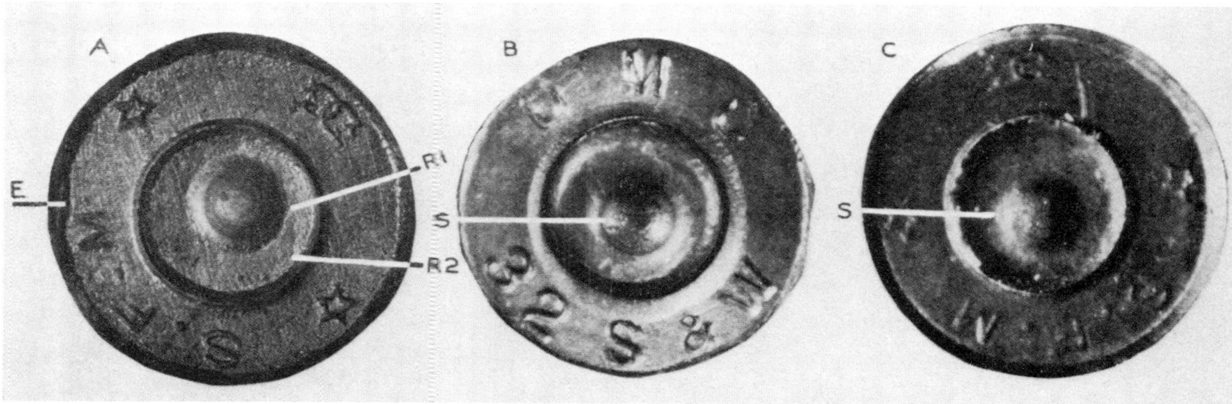

THE CASE OF THE SHOT CYPRIOT DOCTOR

A is a cartridge fired from the pistol which was found on the premises of the accused man. The ejector mark is indicated by E, and the ridge mentioned in Plate XXXVII by R1 and R2. This cartridge should be compared with the six cartridges in Plate XXXVII when it will be found that the markings correspond. B and C are the two cartridges fired on the scene of the crime by the murderer. In neither of these cartridges is there any sign of an ejector mark, nor the ridge which was imprinted by the deep tool cut on the suspect pistol. But in both these cartridges there is a small, but distinct, striker scrape indicated by S. No such scrape was present on any one of the cartridges fired for test purposes from the suspect pistol. This striker scrape enables the cases to be oriented to correspond with the orientation of the test cartridge. It will be seen that in both the crime cartridges the "ex-centricity" of the striker indentation is towards six o'clock instead of one o'clock as it was in all the test cartridges

examining the breech of the suspect pistol I found that, sure enough, there was a deep tool cut across the face on the opposite side of the striker hole to the ejector.

I was by now convinced that there had been a mistake and that the pistol found hidden in Petrou's cellar could not possibly have been used for the murder of Zemenides, but I was determined to make every check that was possible, and so turned my attention to the "crime" bullet.

Here the difference was, if possible, even more pronounced. For instead of being lightly engraved by an obviously loose barrel, the bullet was heavily engraved all round both by the lands and grooves of what was clearly a very tight barrel, and the land engraving was so deep that it reached well up above the cannelure.

There was no need for any examination of striations or minor markings: the difference in the major markings was glaring.

But even now I felt that more evidence would be required to convince a non-technical jury in view of the damning chain of circumstantial evidence against the wretched Petrou. So Mr. Wilson and I decided to repeat our firing test, using pistol ammunition of the S.F.M. brand, which was the brand used for the fatal shot, and ·32 revolver ammunition with the rims filed down. Our purpose was to provide evidence which would satisfy a jury that the thumb-mark of the pistol *must* be imprinted on every cartridge fired, irrespective of its make; and we felt that the greater the number of fired cartridges that we could produce with this thumb-mark, the more convincing would be the testimony in the eyes of a jury.

N

Accordingly we repeated the test with 48 rounds of S.F.M. pistol ammunition and 13 rounds of ·32 revolver ammunition with the rims filed down. The S.F.M. ammunition was very old, and many of the pressures were very feeble—far weaker than those which had been developed in either of the crime cartridges, as could be told by the striker indentations: the pressure developed in the revolver cartridge with the filed rim had been normal, while that in the pistol cartridge "soft," although far from feeble.

Consequently this second test of ours was very severe. But the result was the same, and on every single cartridge case of the whole 110 which we had fired from the suspect pistol the same thumb-mark was visible.

On hearing our opinion the defending solicitor spared no pains in making the most searching enquiries, the result of which convinced him that the whole affair was a "plant." The doctor had had many enemies in the Cypriot colony, and it seemed probable that one of these had actually shot the doctor and had fled the country later on.

Petrou's story had, from the very first, been a complete denial of everything. He declared that he had never even seen the pistol before, let alone bought it, and that he knew nothing of the murder. He merely reiterated "I am innocent and God will help me."

Accordingly the defence took the line that the whole case against Petrou was a "frame up," and that the pistol found in the cellar of his lodgings had been hidden there by one of the conspirators in order to throw suspicion on Petrou. Unfortunately for the conspirators they had planted the wrong pistol.

The most dangerous point was the similarity between

the ammunition found in the pistol and that of the "crime" cartridges. But, as I have explained before, it is by no means easy for unauthorised persons to obtain ammunition, and so filing the rims off revolver cartridges is a common practice in order to make them suitable for self-loading pistol.

We were very pressed for time, but I took as many photographs as I could of different test cartridge cases, and I selected two test bullets of exactly the same diameter as the "crime" bullet and took twelve photographs of each of these, as well as of the "crime" bullet. The two tests bullets were entirely free of any sign of groove engraving, although of identical diameter to the "crime" bullet which was heavily engraved all round.

Plate XXXVII shows six of the test cartridges, two F.N., two S.F.M. and two ·32 revolver cartridges with the rims filed. In every one the ejector mark and "ridge" are clearly shown, while it will be seen that with the ejector mark at 9 o'clock the ex-centricity of the striker indentation is constantly towards 1 o'clock.

Plate XXXVIII shows another S.F.M. test cartridge, which I have included for purposes of comparison, and the two "crime" cartridges.

The photographs of these two "crime" cartridges are not as clear as they should be. This was because I was given to understand that I would have to photograph the Crown exhibits in London, where the only available place was my office.

I was, therefore, compelled to improvise a portable apparatus which could not possibly be as steady as a proper bench. I was also greatly troubled by traffic and other vibration, and even though I reduced my exposures by omitting a screen, there was enough vibration slightly to blur the image, while the omission of the screen resulted in a certain loss of contrast. But difficulties such as these are typical of those under which the defence may have to work, although it is a pleasure to be able to place on record the ungrudging help which was given wholeheartedly by the police themselves. In fact, I have always found the police only too ready to help in every way, and it has invariably been a privilege to see how anxious they are to be absolutely fair.

The photographs of the two "crime" cases, however, show that there was no ejector mark, nor any "ridge" on either, but in each case the end of the line indicates the small striker scrape. Not only did this convince me that the pistol which had been used for the murder probably had an abnormally long striker or was of a type in which the striker acted as the ejector—for there was no sign of a striker scrape in any one of the 110 test cartridges—but, as I have said, it enabled me to orientate the "crime" cases as if they had had ejector marks. And when orientated in a manner corresponding to the test cases, as has been done in Plate XXXVIII, the striker indentation is in each case out of the centre towards 6 o'clock. The bases of both the "crime" cartridges were scored by the lip of the magazine, as described on page 129, but those marks were of no moment. And the absence of the ejector mark suggested that the pistol must have been of a type in which either the firing pin, or the next cartridge in the magazine acted as an ejector.

Plate XXXIX shows four different views of the two test bullets, and Plate XL four views of test bullet No. 2

THE CASE OF THE SHOT CYPRIOT DOCTOR

The upper row gives four different views of a bullet fired for test purposes from the pistol which was found on the premises of the accused man. The lower row gives four different views of a second test bullet fired from this same pistol. The diameters of both these two test bullets are identical. It will be seen that they fitted the bore very loosely, as the land engraving is shallow and barely reaches beyond the top of the cannelure, while neither bullet bottomed any of the grooves. The similarity between the major markings of the rifling engraving on these two bullets is very pronounced

and four similar views of the " crime " bullet. When it is remembered that the mean diameters of all these three bullets were identical, the difference in the engraving of the rifling is blatant.

The Home Office expert based his evidence of identification chiefly on bullet evidence. He had fired a single test round, and cross-examination elicited the fact that he had examined no pistol other than the suspect weapon. He declared that the markings on the base of the test cartridge were identical to those on the two " crime " cartridges, but this statement was supported by no evidence other than photographs of the " crime " cartridges. This absence of a similar photograph of his test cartridge, which was essential for purposes of comparison, obviously rendered this evidence of little value. I examined his test cartridge in court and the ejector mark and " ridge " could easily be distinguished with a pocket lens.

His test bullet was engraved by the grooves on one side, but it was markedly larger in diameter than the " crime " bullet which, as I have stated, was engraved all round.

His evidence of identification was supported by photomicrographs of the land engravings of the " crime " bullet and his test bullet. These photographs were taken at a high power, and the direction of illumination was different in the photographs of the " crime " bullet from what it was in those of the test bullet. This was obvious from the fact that in one series of photographs the top side of the cannelure was brilliantly illuminated, while in the other set this same side of the cannelure was in deep shade. And even then the striations in the two sets of land engraving were quite different in the two sets of photographs, while owing to the high power at which the photographs were taken they were by no means easy for a non-technical observer to interpret.

The jury returned a verdict of Not Guilty.

CHAPTER IX

THE IDENTIFICATION OF THE MAKE OF FIREARM

THERE is one problem of identification which I have not as yet mentioned.

On the discovery of a "crime" cartridge case, or the recovery of a "crime" bullet, the question may arise as to whether it is possible to ascertain the particular type and make of weapon with which the crime was committed. And since a "crime" bullet is probably recovered more frequently than a "crime" cartridge case I will try first to consider the problem as presented by a fired bullet.

It has already been seen that rifling varies in the number of grooves, relative sizes of grooves and lands, direction of twist, depth of groove, and pitch. So if data is available giving these details of rifling for all makes of pistols and revolvers, it is theoretically possible to determine the particular make of weapon which fired a particular bullet by an examination of the engraving.

Great attention has been paid to this work in America, but my own private and personal opinion is that the work is of little practical value, at any rate in England.

My reasons for holding this view are three:

(1) There is a possibility of both human and instrumental error in reading the engraving on a fired bullet,

and the error which thus arises may easily assign the bullet to an incorrect make of weapon.

(2) There is no constancy in specifications used in the rifling of large numbers of the cheaper continental pistols, and so it is useless to bother about minute measurements when there are much bigger variations in the actual rifling.

(3) Neither the police nor the public have a sufficient knowledge of firearms for the information, even if correct, to be of any real value.

Let me try to explain these reasons at greater length.

Bullets do not always bottom the grooves, and consequently it is then impossible to measure the depth of the groove by measurements of the engraving on the bullet. But in any case all measurements made with a micrometer eye-piece are difficult to make and require much skill and practice. The difference in the width of lands in some makes of weapons is so slight that I doubt if it is possible to measure the difference by means of the engraving on a fired bullet with sufficient certainty to be able to say that the bullet must have been fired by one make and not by another make.

The most distinct source of variation in different systems of rifling is the pitch, and measurement of the pitch is certainly the best way of obtaining results. But pitch is not easy to measure. The Americans use a gonimeter eye-piece, which is really a type of eye-piece for measuring small angles through a microscope. But not only is such an eye-piece difficult to use, but the actual striations on a fired bullet frequently vary in direction in a most disreputable manner owing to the "skid marking." It is certainly a fact that the

THE CASE OF THE SHOT CYPRIOT DOCTOR

The upper row gives four different views of the fatal bullet. This bullet must clearly have been fired through a very tight barrel, as the land engraving is very deep and heavily scored and reaches well above the top of the cannelure, while the striations caused by the groove engraving extend all round the circumference of the bullet. The lower row shows four similar views of one of the test bullets which was fired from the suspect pistol. This bullet is of exactly the same mean diameter as the fatal bullet, but the difference in the rifling engraving is obvious

majority of the striations will be constant in direction, but even then it is in practice by no means easy to obtain a good datum line which is truly parallel to the major axis of the bullet.

Then bullets do not always travel absolutely nose-on down the bore, and when this happens the engraving is cut at a false angle.

So, taking one thing and another, I doubt whether it is possible always to be *certain* of reading to within 5 per cent. of the true pitch with a goniometer eye-piece, and unless the observer has exceptional skill the error may be much greater.

My own view is that in the preparation of any sort of evidence everything possible should be done to eliminate the human element. An expert witness in firearms identification cases should produce facts, and not opinions which cannot be checked. He should be able to say: "Here are my photographs: these are the obvious points of similarity or difference: judge for yourself." His skill and knowledge will be tested by obtaining the photographs.

It may be urged that such preliminary investigations have nothing to do with evidence, and this is true. But the principle remains, and I cannot but regard it as a risk even to begin to adopt methods which are dependent solely on the individual skill of a particular observer in making measurements through the microscope. There is no harm in such work being carried out provided it is never brought forward as evidence later on.

But my second reason should be more convincing. It is undoubtedly true that the manufacturers of high-grade weapons, such as Webley, Colt and Smith &

Wesson, do rifle their arms accurately to specification. But various continental manufacturers of cheap pistols, which are much more common than high-class weapons, work under no such handicap. There can be no doubt but that they purchase lengths of barrel already rifled, and cut these lengths up into pistol barrels. But they do not always buy these lengths from the same factories, and consequently there is no sort of standardisation. My friend Mr. Wilson has in his possession two ·32 self-loading pistols of the same make, one of which is rifled with a right-hand twist and one with a left-hand twist! When anyone is trying to determine the type of weapon which fired a particular bullet he is assuming a standardisation in manufacture which does not exist in many of the more common makes of weapon.

Finally I cannot help doubting the practical help to the police in telling them that a certain fatal bullet was fired by, say, a Victoria self-loading pistol so long as the majority of police officers in this country do not know the difference between a self-loading pistol and a revolver! Quite recently I heard an exceptionally efficient and able police sergeant say that he had found a revolver among the effects of an injured man. I immediately offered to bet his Chief Constable that the weapon would turn out to be a self-loading pistol. Unfortunately the Chief Constable was so exercised as to what I meant by a self-loading pistol that he had not time to take my bet before the weapon was produced. This was a pity as I would have won my bet!

Frankly I cannot see how any criminal investigation can be helped by the knowledge that the weapon used by an unknown criminal must have been of some particular make. It is a simple matter to examine any suspect weapon that may be forthcoming from time to time, and either reject it at once as being impossible, or else to test it more thoroughly if it is a "possible."

It is because I hold these views that I have not devoted any serious attention to this particular branch of firearms investigation. But on occasions negative evidence of great value may be obtained from engraving on a fired bullet.

Let me give an example.

A couple of years ago I was called across England at short notice. A young man had been practising in his garden with a ·22 rim fire rifle, and another man who had been working beyond his boundary had been shot. The distance was about 150 yards. The bullet perforated the second man's body from rear to front and was extracted from his abdominal cavity. Quite naturally he claimed damages from the man whom he alleged had shot him.

The first man denied the possibility of his having shot the second, declaring that he had been firing in another direction. So the second man instructed his solicitors to bring an action for damages.

The defendant's solicitors called me in with a view to ascertaining whether the bullet extracted from the plaintiff had, or had not, been fired from their client's rifle.

I was informed that the defendant was using a Walther rifle at the time of the accident. Now all ·22 rifles are rifled with either 6 or 8 grooves. Some may have 7, but the almost universal number is 6 or 8. The Walther, however, has but 4 grooves. If, therefore, the fired bullet was engraved with 6 or 8 grooves the defendant would be completely exonerated; but if it was engraved with 4 grooves his position would not be so secure.

I went to the offices of the plaintiff's solicitors and examined the bullet. It was engraved with but four grooves.

So I told the defendant's solicitors that unless they could produce someone else who was using a Walther rifle at the same time and within, say, a quarter of a mile of the scene of the accident, they had better settle out of court. I added that if they were really anxious for me to do so, I would do my best to see whether the plaintiff's bullet did, or did not, marry the defendant's rifle. But that apart from the time and cost of such work, if the result was in the affirmative their client would be in a worse position than he was already; and the risk seemed hardly worth while. I explained that I could not say definitely that the bullet had been fired from *the* Walther rifle without further tests, but that it had undoubtedly been fired from *a* Walther rifle.

They took my advice and settled out of court.

But if the bullet had been engraved with more than four grooves the defendant would have had a cast-iron case.

I do not wish to appear to belittle the value of examinations of the rifling engraving on fired bullets, because so long as only major characteristics are taken into consideration the information which can be obtained may prove of great help. But I do feel that the examination of the more detailed characteristics may possibly lead to incorrect conclusions partly on account of the practical difficulty of making the examinations, but chiefly because of the complete lack of standardisation which exists in many of the factories in which some of the cheaper varieties of pistols are made.

These views of mine have been endorsed to a very large extent by the German investigators, Doctors Otto Mezger and Walter Hess and Criminal Inspector Fritz Hasslacher, who are the joint authors and compilers of an invaluable monograph on the *Determination of the Type of Pistol employed from an Examination of Fired Bullets and Shells.* This monograph is accompanied by an "Atlas" giving the essential particulars of the types of rifling as well as of the relative positions of the various components of the breech faces of one hundred different makes of self-loading pistols.

The authors are of the opinion that the use of a goniometer eye-piece is not sufficiently reliable for measuring the pitch of the engraving on a fired bullet, and they have devised an apparatus in which they use a stereoscopic microscope with a stage and bullet holder of special design. It is claimed that with this apparatus the rifling pitch can be measured from the engraving on fired bullets, including those which are somewhat deformed, with a degree of accuracy which is considerably greater than that attainable with a goniometer eye-piece.

These German investigators classify self-loading pistols according to their types of rifling, and give details of all the types, but state emphatically that owing to the lack of standardisation in manufacture it is utterly impossible to determine the actual *make* of the pistol used for firing a particular bullet.

Fired cartridge cases present a somewhat different problem. It is easy to tell the calibre, as not only is this usually stamped on the base, but the size of cartridge is unmistakable. And if the cartridge is rimmed it must have been fired from a revolver—assuming that it is not a rifle cartridge. If the cartridge is a rimless self-loading pistol cartridge the probability is that it was fired from a self-loading pistol although, as has been seen, revolvers are made to take pistol ammunition in ·45, ·32 and ·25 calibres. But if there is an ejector mark on the case the assumption that the weapon used was a self-loader becomes overwhelming.

The various makes of self-loading pistols differ in the relative positions of their extractors and ejectors, and consequently if the necessary data is available, it is theoretically possible sometimes to determine the actual make of pistol used. The German " Atlas " to which I have referred provides such data and might be of considerable use to an experienced investigator. But I fear that it might easily lead an inexperienced man astray because he would search for marks shown in the photographs of the "Atlas," but which did not happen to be reproduced on a particular fired case because of a low pressure or some other reason. I have found, for instance, one particular model of a particular make of self-loader which failed entirely to produce the theoretically essential characteristic mark of that particular make.

So, as with fired bullets, I am inclined to the view that it is impossible to be certain of determining the make of pistol used from an examination of a fired cartridge case, although at times it may be possible to narrow the issue down to a few particular makes.

With revolvers there is no possible means of telling the make from examining a fired cartridge case, and even the German investigators do not seem to have tried.

I can but repeat, therefore, that my own purely personal opinion is that it is waste of time and effort to try to do much more than determine the general type of weapon used by examinations of either fired bullets, or cartridge cases, or both. And that the better plan is to examine any suspect weapon that may be forthcoming from time to time when "impossibles" and "possibles" can be ascertained quickly and with absolute certainty. The "possibles" should be used for firing a few test rounds when the procedure which I have suggested in the previous chapters will give results which are definitely positive or negative.

INDEX

A

B

C

D

E

F

I

J

K

L

M

P

R